Current Perspectives in Human Physiology: Selected Readings

Edited by

Linda Vona-Davis
West Virginia University

Lauralee Sherwood
West Virginia University

West Publishing Company
Minneapolis/St. Paul New York Los Angeles San Francisco

WEST'S COMMITMENT TO THE ENVIRONMENT

In 1906, West Publishing Company began recycling materials left over from the production of books. This began a tradition of efficient and responsible use of resources. Today, 100% of our legal bound volumes are printed on acid-free, recycled paper consisting of 50% new paper pulp and 50% paper that has undergone a de-inking process. We also use vegetable-based inks to print all of our books. West recycles nearly 27,700,000 pounds of scrap paper annually—the equivalent of 229,300 trees. Since the 1960s, West has devised ways to capture and recycle waste inks, solvents, oils, and vapors created in the printing process. We also recycle plastics of all kinds, wood, glass, corrugated cardboard, and batteries, and have eliminated the use of polystyrene book packaging. We at West are proud of the longevity and the scope of our commitment to the environment.

West pocket parts and advance sheets are printed on recyclable paper and can be collected and recycled with newspapers. Staples do not have to be removed. Bound volumes can be recycled after removing the cover.

Cover Image: Immunofluorescent image of rat fibroblasts in which vaults (red) are clustered on actin stress fibers (green). Nancy Kedersha, ImmunoGen.

Production, Prepress, Printing and Binding by West Publishing Company.

Printed with Printwise
Environmentally Advanced Water Washable Ink

Printed in the United States of America
03 02 01 00 99 98 97 96 8 7 6 5 4 3 2 1 0

ISBN 0–314–07564–X

TABLE OF CONTENTS

Part Three: The Brain, The Peripheral Nervous System and Properties of Neurons That Serve Them Both

Part Four: The Mechanical Forces and Chemistry of the Senses

Part Seven: Controversies and Ethical Issues

Preface

Nowhere in the world of science is there more relevance to our everyday lives than in the study of physiology. Many of the concepts of physiology were developed from the fundamental principles of biology, chemistry and physics and its diversity is reflected in the wide variety of research reports published in scientific journals. Thus, it becomes a lifelong challenge for teachers and their students to advance their understanding of physiology by reviewing the current literature.

This collection of newsworthy articles provides a more readable and informative approach for students to supplement the material often encountered in the introductory study of physiology. These articles were carefully selected from general interest periodicals and science magazines and organized into topics of special interest to the student. Each article begins with a brief summary of the ideas presented in the selection and ends with a set of questions to help identify the key points of discussion.

Acknowledgments

The editors and West Publishing wish to express their sincere thanks to the many magazines that allowed us to reprint their articles. We are indebted to Crystal Denette, who typed and produced the camera-ready pages for this book. Her round-the-clock work helped make this book as up-to-date as possible.

Linda Vona-Davis
Lauralee Sherwood

Part One:

Life in the Cell

Nerve cells need a constant supply of synaptic vesicles to package the neurotransmitters that are needed in signal transmission. To accomplish this feat, the plasma membrane forms a pit lined with the protein clathrin beginning the process of endocytosis. Once the pit is completely formed into a sphere, the vesicle is severed from the membrane by a helical structure called dynamin which wraps around the neck of the vesicle pinching it off from the membrane. Scientists studying this unique protein have found this large GTPase to be necessary for the endocytosis of surface membranes in a variety of cell types, including neurons. Of particular interest is the role of dynamin in protein sorting and membrane fission using these helical structures.

Ringing Necks with Dynamin

by Regis B. Kelly

Synaptic vesicles in nerve cells are packages of neurotransmitters that are ejected each time a neuron fires. The stock of vesicles is then replenished rapidly by endocytosis of their membrane components from the plasma membrane and repackaging with neurotransmitter once inside the cell. An agent instrumental in sustaining the supply of vesicles, dynamin, has now been collared.[1,2]

Dynamin, a large GTPase, is needed for the endocytosis of surface membrane in a large number of cell types, including neurons.[3] Hinshaw and Schmid[2] show that it can adopt a helical structure in solution; Takei *et al.*[1] demonstrate that it is this helical structure that enables dynamin to wrap around the neck of membrane vesicles and help pinch them off from the plasma membrane. Dynamin-like molecules do not seem to be universally required for membrane fission. An ability to self-assemble into helical arrays, however, may be a common feature of dynamin and its homologues and could explain their physiological properties.

Endocytosis of membrane proteins from the plasma membrane involves three steps[4]: clustering of the membrane proteins by clathrin and its associated proteins, invagination of the membrane into a constricted clathrin-coated pit and, finally, severing of the neck to give a coated vesicle. *In vitro*, this sequence can be arrested at different stages, depending on which analogue of GTP is present[5]. The defining characteristic of the constricted-vesicle stage is that the neck allows small molecules to pass into the vesicle, but excludes protein. Structures that look like constricted vesicles are found at the surface of mammalian and *Drosophila* cells in the presence of mutant forms of dynamin[6-8]. Although it seemed plausible that GTP hydrolysis by dynamin could be necessary for severing the neck of the coated pit, it was not clear just how it did it.

Dynamin is abundant in brain (1.5 per cent of total protein), and possesses, in addition to a GTPase domain, a pleckstrin-homology domain, and a carboxy-terminal proline-rich domain containing an SH3-binding motif[3]. The demonstration of its ability to self-assemble into long helical structures in the test tube[2] is complemented by the elegant electron micrographs from DeCamilli's laboratory[1], which show that dynamin surrounds the neck of the membrane. When permeabilzed nerve terminals are incubated in the presence of the non-hydrolysable GTP analogue, GTP-γS, long invaginations are formed which are coated with spirals of dynamin. Clathrin-coated bulbs are often seen at the tips of the invaginations. The spiral of polymerized dynamin squeezes the membrane tube to an outside diameter of about 30 nm, and an internal diameter tight enough to prevent the passage of proteins. The GTP-associated form of dynamin is believed to ring the neck of the constricted pit[1,2]; GTP hydrolysis could change the configuration of the helix, squeezing the neck further until membrane fission occurs in a sort of molecular garroting.

This pinching-off of membrane necks occurs frequently in biology, in events as diverse as cell division, budding of membrane-coated viruses, and vesicle budding from an intracellular organelle. If a molecule like dynamin is required in each case, then GTP hydrolysis should be necessary for membrane fission. This is not true for budding from the endoplasmic reticulum or Golgi, where GTP hydrolysis is required for vesicle uncoating and docking but not for vesicle formation. Dynamin may be able to help bud small vesicles from the plasma membrane but not from intracellular membranes because of the difficulty of curving a stiffer membrane. Plasma membrane differs fundamentally from intracellular membranes in that it is supported on the inside by cortical cytoskeleton and is attached to matrix and other cells on the outside.

Homologues of dynamin all have highly similar GTPase domains. One class of analogues, the Mx proteins, can confer resistance to viral infection. Until now it has been difficult to see a functional link between dynamin and homologues that regulate viral infection. An exciting clue comes from the observation that the mouse Mx protein can also polymerize into C-shaped and helical molecules, depending on the nucleotide concentration, that are 11 nm thick and 100-150 nm long, parameters similar to those found for dynamin-1. Thus the dynamin family may share a propensity to form filamentous aggregates, regulated like mictotubules by GTP hydrolysis. Perhaps Mx proteins interfere with virus infection by wrapping around cylindrical structures of about 25 nm diameter.

There is also an intriguing link between dynamin and VPS1 protein, a dynamin-like protein in yeast involved in protein sorting. Mutations in VPS1 send a vacuolar proteases to the surface and a Golgi enzyme to the vacuole[10,11]. Most models of protein sorting require a mechanism for removing membrane proteins from an organelle, leaving behind the soluble protein content. Such segregation is needed, for example, to return low-density lipoprotein (LDL) receptors from the endosome to the cell surface, after the LDL particles themselves have been internalized for the purposes of removing their cholesterol cargo. The canonical model is that a membrane protrusion forms, thereby excluding soluble proteins. Considering dynamin's remarkable ability to constrict tubes, it is feasible that a dynamin analogue such as the VPS1 protein could facilitate membrane protrusion (and therefore membrane sorting) by a common mechanism. The VPS1 protein and dynamin-1 both have the 'self-assembly' domain essential for the polymerization of mouse Mx protein[9], suggesting that VPS1 may also form helices. Hinshaw and Schmid's data, however, caution us that the proline-rich, SH3-binding, carboxy-terminal region, which is not conserved among dynamin homologues, is required for polymerization of dynamin *in vitro*.

Nonetheless, the roles of dynamin in membrane fission and of VPS1 in sorting probably both depend on the ability to constrict membrane tubes using helical polymers. If this turns out to be the case, then other mutations that affect sorting should involve dynamin-like proteins. ❑

References

1. Takei, K., McPherson, P. S., Schmid, S. L. & De Camilli, P. *Nature* **374**, 186-190 (1995)
2. Hinshaw, J. E. & Schmid, S. L. *Nature* **374**, 190-192 (1995)
3. Vallee, R. B. & Shpetner, H. S. *Nature* **365**, 107-108 (1993)
4. Robinson, M. S. *Curr. Opin. Cell Biol.* **6,** 538-544 (1994)
5. Carter, L. L., Redelmeier, T. E., Woollenweber, L. A. & Schmid, S. L., *J. Cell Biol.* **120,** 37-45 (1993)
6. Kosaka, T. & Ikeda, K. *J. Neurobiol.* **14,** 207-225 (1983)
7. Koenig, J. H. & Ikeda, K. *J. Neurosci.* **9,** 3844-3860 (1989)
8. Damke, H., Baba, T., Warnock, D. E. & Schmid, S. L. *J. Cell Biol.* **127**, 915-934 (1994)
9. Nakayama, M. *et al. J. Biol. Chem.* **268,** 15033-15038 (1993).
10. Vater, C. A., Raymond, C. K. Ekena, K., Howald-Stevenson, I. & Stevens, T. H. *J. Cell Biol.* **119,** 773-786 (1992)
11. Wilsbach, K. & Payne, G. S. *EMBO J.* **12,** 3049-3059 (1993)

****

Questions:

1. What are synaptic vesicles?

2. What is endocytosis?

3. What role does the protein dynamin play in endocytosis?

Answers are at the end of the book.

What happens when the body's own immune system doesn't recognize it? The result is an autoimmune disease whereby war is raged by T cells on the very tissues they should be defending. To weaken or suppress the activation of harmful T cells, scientists in many laboratories have begun testing a phenomenon called oral tolerance. The idea is to administer synthetic proteins to the immune system that match crucial proteins in the body, thus restraining the immune response by teaching it a lesson in tolerance. It is a technique that shows tremendous potential for the treatment of multiple sclerosis, rheumatoid arthritis and to prevent the rejection of transplanted organs.

Teaching Tolerance to T Cells

Sometimes our own immune system attacks crucial proteins in our bodies, and we get sick. A possible solution: Eat those proteins.

by Sarah Richardson

Autoimmune Diseases result from a cruel mistake: the patient's immune defenders, notably the T cells, attack the very tissues they should protect. In rheumatoid arthritis, it's the joints; in diabetes, the pancreas; and in multiple sclerosis, the brain and spinal cord. For some reason the immune cells botch their central mission, which is to distinguish self from nonself. Now researchers at a number of institutions are testing a novel approach to restraining the traitorous T cells. It's based on a simple principle: You are what you eat.

When you eat food, you ingest proteins that are unfamiliar to the body. But immune cells in the gut don't attack those proteins; they allow them to be absorbed. Because we need to take in food, the gut's immune cells have evolved to be more tolerant of foreign proteins. Furthermore, they apparently communicate this tolerance to immune cells elsewhere in the body. If you inject a forgein protein into a person or an animal, you normally get a strong immune response. But if a small amount of the protein is eaten before or after the injection, the immune response is weaker. The phenomenon is called oral tolerance.

In the 1980s neurologist Howard Weiner of Harvard Medical School was one of the first researchers to recognize the potential of oral tolerance as a treatment for autoimmune disease. "As I was reading and thinking," says Weiner, "I said to myself, wait a minute, we don't normally react to proteins that come into the gut, because of the way we evolved. But in multiple sclerosis, we're reacting to a protein in the brain"—a "self" protein. "Could we reset the immune system? What if I fed MS patients a brain protein? Would it help?"

Weiner and his colleagues first tested this idea by injecting mice with one of the proteins that make up myelin—the material that forms a protective sheath around nerve fibers and that in multiple sclerosis is attacked by T cells. After the injections, the mice contracted the mouse equivalent of MS. But when Weiner gave the animals the myelin protein to eat, the number of myelin-attacking T cells in their bodies decreased substantially.

Why? When the gut absorbs myelin, says Weiner, regulatory T cells are released that migrate out of the gut and are distributed through the bloodstream. "It's like an immunization in the gut, where you induce these regulatory immune cells that then go to the brain and suppress inflammation," he explains. Although the regulatory T cells appear to be structurally identical to T cells that myelin, they don't secrete the same chemical signals. Instead of releasing signals that escalate the immune attack, they release signals that suppress the activation of harmful T cells.

More recently, Weiner and his colleagues conducted a yearlong study of 30 human MS patients. Half were fed myelin pills every day, while the other half got a placebo. Those fed the myelin suffered fewer of the abrupt neurological attacks that characterize MS (the cumulative effect of which is a degeneration of the legs, eyes, and other parts of the body). A different research group is now testing the treatment in a full-scale clinical trial involving 504 MS patients.

Clinical trials of oral tolerance are also under way on patients with rheumatoid arthritis and uveitis, an autoimmune disease of the eyes, and a trial is planned for juvenile diabetes. One day oral tolerance may even be used to prevent the rejection of transplanted organs. Before receiving the donor organ, the recipient would eat synthetic proteins that match crucial proteins in the organ, thus blunting the immune response. This idea has been successfully tested in animals.

Oral tolerance is not likely, however, to become a cure for autoimmune diseases; it merely restrains the autoimmune response, rather than eliminating whatever causes the response. And a lot remains to be learned about how oral tolerance works. But Weiner thinks it is already clear that oral tolerance can yield effective treatments—and without the side effects that accompany some of the drugs that are now used to treat autoimmune disease. "The amazing thing is, it's not toxic," he says. "It stimulates the body's own regulatory systems."

"Ten years ago," he adds, "there weren't many people working on this approach. Now they realize how powerful a mechanism oral tolerance is." ❑

****

Questions:

1. What is an autoimmune disease?

2. How does the body trick the T cells in oral tolerance?

3. Is it possible to cure autoimmune diseases with oral tolerance?

Answers are at the end of the book.

Multiple sclerosis (MS) is a pathophysiological condition which causes demyelination of nerve fibers in various locations throughout the central nervous system. Loss of the myelin exposes the axonal membrane to the extracellular fluid which slows transmission of impulses. Since it is an autoimmune disease, scientists are searching for autoantigens which might become induced in the diseased state. Recently, a small heat-shock protein called αB-crystallin was identified as the myelin protein target of the immune response in MS. New strategies can now be developed to promote the induction of immunological tolerance to this protein and others proteins in patients suffering from this debilitating disease.

Presenting an Odd Autoantigen

by Lawrence Steinman

Multiple sclerosis (MS) is an autoimmune disease involving an integrated attack by T cells, B cells and macrophages on the myelin sheath that surrounds nerve fibres. Potentially inflammatory cytokines such as γ-interferon and tumour-necrosis factor-α are found at the sites of damage, and T cells and antibodies directed against myelin components can be isolated from inflamed regions in the central nervous system. Attempts to identify the components of the myelin sheath that provide this misguided arsenal have yielded a number of candidates. Van Noort *et al.*[1] describe the isolation of a protein that is a prominent target at the disease site in the myelin sheath: this autoantigen turns out to be a small heat-shock protein, αB-crystallin, which is induced in the diseased white matter.

Van Noort and colleagues reasoned that there could be an antigen present in the white matter of MS brain that is not found in normal white matter and which specifically activates T cells. They separated the proteins of the myelin sheath using reversed-phase high-performance liquid chromatography and discovered that a particular fraction in the myelin of MS brain, but not in the myelin taken from healthy brain, stimulated proliferation of T cells, causing them to release the proinflammatory cytokines γ-interferon and interleukin-2. αB-crystallin was identified as the myelin protein that elicited the strongest immune response. They then showed that αB-crystallin is expressed in glial cells from MS lesions but not in white matter from healthy individuals or in unaffected white matter from MS brain. αB-crystallin was found in oligodendroglial cells as well as in astrocytes in plaques from patients with acute and chronic MS.

These exciting results provide a strong case for αB-crystallin as a target of the immune response in MS, but they raise the question of the precise role of αB-crystallin in the pathogenesis of the disease. αB-crystallin is constitutively expressed in the lens of the eye, myocardial cells and kidney epithelium, and is inducible not only in MS tissue, but also in brains of patients with other neurological diseases, including Alzheimer's, Parkinson's and Huntington's diseases.[1] Degenerative diseases like Huntington's and Parkinson's do not appear to be medicated by the immune system, so it is not clear why αB-crystallin should be immunogenic in MS but not in these other degenerative conditions. One possibility is that the induced expression of αB-crystallin may be a reaction to some neuropathological stimulus that is not unique to MS. In MS additional local events may be required to allow the development of an immune response to αB-crystallin.

To assess the significance of the immune reaction against αB-crystallin in the pathology of MS, the immune response to other myelin components in MS needs to be considered. Myelin basic protein, proteolipid protein, transaldolase and 2',3'-cyclic nucleotide 3'-phosphodiesterases, as well as two members of the immunoglobulin supergene family found in the myelin sheath (myelin oligodendroglial glycoprotein and myelin-associated glycoprotein) all trigger immune responses in MS patients[2]. When they are injected into laboratory animals, these myelin components cause allergic encephalomyelitis, with inflammation and demyelination in the central nervous system. In addition, other inducible heat-shock proteins can be detected in glial cells in MS lesions and can stimulate an immune response in MS patients.[3-5] The large number of antigens capable of eliciting an immune response in MS patients may represent the intermolecular dispersion of an immune response that arose initially against a single component of myelin. Immune reaction to different components of a supramolecular structure, such as the myelin sheath in MS or the mitochondrion in primary biliary cirrhosis, is common in individuals with autoimmune disease that involves a discrete organ.

We cannot yet say which of these antigens is most important in the pathogenesis of MS. Strong evidence is emerging that an immune response to certain regions of myelin basic protein and proteolipid protein could be critical in MS. The major T- and B-cell response in the central nervous system of MS patients who carry the human leukocyte antigen HLA-DR2 (about two-thirds of patients) is directed to a region between residues 84 and 103 of myelin basic protein.[6,7] The principal antibody response in the lesions of MS patients is also directed against this region[8] and microbes that share critical sequence homologies with it can provide a response to this myelin peptide.[9] Increased numbers of T cells reactive to proteolipid protein are also found in the spinal fluid of patients with MS.[7] The response to αB-crystallin has not yet been checked for T and B cells in the brain.

These efforts to determine which antigens trigger the pathological response in MS brain have very practical consequences. It is now possible to induce immunological tolerance to specific proteins using a variety of strategies, including the alteration of peptide ligands that bind to the T-cell receptor and the blockade of costimulatory molecules on T cells.[2,10] All of these approaches have been effective in suppressing animal models of MS. Identification of the critical proteins involved in MS permits the testing of these strategies for the induction of tolerance in patients suffering from this perplexing disease. ❑

References

1. van Noort, J.M. *et al, Nature* **375**, 798-801 (1995).
2. Steinman, L., Waisman, A. & Altmann, D. *Molec. Med. Today* **1**, 79-83 (1995).
3. Selmaj, K., Brosnan, C. & Raine, C.S. *Proc. Natn. Acad. Sci. U.S.A.* **88**, 6452-6456 (1991).
4. Wucherpfennig, K.W. *et al. Proc. Natn. Acad. Sci. U.S.A.* **89**, 4588-4592 (1992).
5. Hvas, J. *et al, J. Neuroimmun.* **46**, 225-234 (1993).
6. Oksenberg, J. *et al, Nature* **362**, 68-70 (1993).
7. Zhang, J., *et al, J. Exp. Med.* **179**, 973-984 (1994).
8. Warren, K., Catz, I., Johnson, E. & Mielke, B. *Ann. Neurol.* **38**, 280-289 (1994).
9. Wucherpfennig, K. & Strominger, J., *Cell* **80**, 695-705 (1995).
10. Weiner, H. W., *et al. A. Rev. Immun.* **12**, 809-837 (1994).

****

Questions:

1. Why is multiple sclerosis considered an autoimmune disease?
2. Where is the αB-crystallin protein expressed?
3. How will this protein be useful in the treatment of MS?

Answers are at the end of the book.

It's a dirty job but it has to be done. Housekeeping proteins called proteasomes help to dispose of old and damaged proteins that tend to accumulate within the cell over time. Described by structural biologists as a hollow, barrel-shaped protein itself, the proteasome is designed to accept proteineious waste into its chambers for grinding into short peptides. How the proteasome knows exactly which proteins should be marked in order to gain entry for disposal is still a mystery. One fact remains certain and that is the necessity of keeping up with the steady stream of molecular garbage that must be removed from every cell.

Garbage Trucks of the Cell

by Josie Glausiusz

Most people pay little heed to garbage collectors until they go on strike and mounds of garbage choke the streets. Likewise, living cells that neglect to grind up their own molecular garbage — pay the price, dying swiftly, mired in their own waste. To cope with such a stream, a cell has tiny garbage trucks, called proteasomes, that roam the cell sucking in waste proteins and breaking them down into small pieces. Although proteasomes have been known about for some time, just how they distinguish between spent proteins and valuable ones has been a mystery. Now, using proteasomes isolated from a bacterium, researchers in Germany have shed light on the process.

A protein consists of a string of amino acids that, in order to function as an enzyme, a hormone, or whatever, must be folded into a complex three-dimensional shape. Wolfgang Baumeister and Thorsten Wenzel, two structural biologists at the Max Planck Institute for Biochemistry in Martinsried, have found that only proteins that have been unfolded back into a simple string get ground up by a proteasome — because only they can fit inside it.

A proteasome is a hollow barrel made of four stacked rings, with entrances at each end. Baumeister and Wenzel designed a unique knotted peptide — a chain of amino acids — to see if it could thread its way through this barrel. The know consisted of a tiny gold particle attached to the peptide. Normally the peptide would have been rapidly broken down inside the proteasome. With the gold particle attached, however, the peptide — as evidenced by an electron micrograph — became stuck at the tunnel's entrance, the gold knot effectively plugging the proteasome's opening.

What is it that unfolds the proteins and enables them to enter the proteasome's narrow channel? It's not entirely clear, but specialized proteins that are attached to the ends of the proteasome, in position to guard the entrances, are apparently involved, according to Baumeister. These guardians seem to selectively unfold the proteins that have been tagged with another protein called ubiquitin. How the cell decides which proteins to mark for destruction in this way, however, remains a mystery. "The whole reason for building such a very complex structure is that the cell must be protected from random degradation of proteins," says Baumeister. "The proteasome has developed that rather narrow channel, and that is protection for the cell. You don't want to have a large hole where anything can leak through."

Once inside the proteasome, the unfolded proteins are broken down into short peptides consisting of between 4 and 11 amino acids. But the researchers don't yet know how the processed peptides exit the molecular garbage mill. "We have no idea how long a peptide remains inside," says Baumeister. "We are thinking about experiments to measure that. And we don't know how they are discharged from the proteasome at all." ❑

****

Questions:

1. What is a proteasome?
2. What does the proteasome do with protein waste?
3. How does the proteasome know which proteins to trap for disposal?

Answers are at the end of the book.

Caveolae may sound like an Italian pasta dish but to cell biologists it means tiny caves. Scientists have been viewing these small indentations on the surface of cells since the 1950s with very few clues to what role they play in the cell. Interest in caveolae is growing now because of the belief that these tiny vesicles have multiple functions, perhaps within the same cell. Its most famous role is in receptor-mediated endocytosis, a process in which the cell carries proteins and other large molecules into the cytoplasm. Other functions include transcytosis, a transport mechanism to escort substances across the cell and possibly signal transduction, a communication system. With all this and possibly more, cell biologists are speculating that defects in caveolae may be a link to cancer and other illness.

Cell Biologists Explore 'Tiny Caves'

Recent work suggests that the small membranous structures called caveolae are specially designed for getting molecules into the cell and are possibly also centers for signal transduction

by John Travis

Any vehicle small enough to drive over the cell's outer membrane would surely have a bumpy trip. The cell's surface is hardly smooth and featureless like an interstate highway. In fact, it's more a gravel-strewn back road—studded with pits and indentations. But unlike potholes, the cellular indentations are far from nuisances. Take the type called 'coated pits" because their inner surfaces are covered by a dense layer of the protein clathrin. These coated potholes play an essential role in receptor-mediated endocytosis, a process in which the cell carries proteins and other large molecules into the cytoplasm.

For many years, most of the attention of biological road crews was focused on the coated pits. Now, though, it's beginning to look as if another, much less studied, membrane indentation—the caveolae ("tiny caves") —may one day rival the impor tance of their clathrin-coated counterparts. Within the past year or two, researchers have shown that caveolae also draw substances into the cell's interior, although in caveolae's case the transported substances are smaller molecules, such as vitamins. There's also growing evidence that the caves participate in signal transduction, the complicated process by which extracellular stimuli like growth signals are transmitted into the cell. Some of the most enthusiastic biologists speculate that defects in the function of caveolae may even be connected to cancer and other illnesses.

These results are causing a strong upswing of interest. Caveolae "have been seen in the electron microscope for over 40 years and only now are we beginning to understand some of the secrets behind their function. It's very exciting. I think there's a whole new field getting started," says cell biologist Richard Anderson of the University of Texas Southwestern Medical Center in Dallas.

Excitement like that was certainly not on the agenda for caveolae back in the 1950s, when they were observed, as little more than dimples in the cell's outer membrane, and named by Japanese electron microscopist Eichi Yamada. Further microscopy showed that they were more than simple invaginations in the cell. They were in fact formed by small vesicles, around 50 nanometers in diameter, fused to the cell membrane. Those micrographs, however, gave researchers few clues to what caveolae actually do.

The earliest functional hypothesis was proposed by cell biologist George Palade, now at the University of California, San Diego, whose group noticed these structures around the same time as Yamada. Because Palade detected caveolae in the endothelial cells lining the walls of blood vessels, he suggested that they might be vehicles for a process called "transcytosis," in which blood-borne molecules are picked up and transported across the endothelial cell to the membrane on the other side where

Reprinted with permission from *Science,* November 19, 1993, Vol. 262, pp. 1208-1209.

their contents can be released into the fluids bathing the cells of the tissues. Subsequent studies showed that radiolabeled molecules that were picked up by the caveolae did, in fact, make their way across endothelial cells as Palade proposed.

Despite such evidence, the transcytosis hypothesis has been controversial for decades. For starters, no electron microscopist could capture a picture of closed caveolae in transit across the cell—which, in fairness to Palade's idea, could have been because they move too fast to be detected or because the tissue preparation methods used in electron microscopy made the structures fuse to the membrane. But there were other uncertainties about caveolae's activities: Their presence in cell types other than endothelium cells suggested that they might have functions beyond transcytosis, but what those functions might be was unclear.

The main problem handicapping research into caveolae was that investigators were limited to looking at micrographs; they hadn't figured out a way to isolate these curious vesicles physically and subject them to rigorous biochemical analysis. Those problems have begun to fade in the 1990s, spurred by what Deborah Brown, a biochemist at the State University of New York (SUNY) at Stony Brook calls the "rediscovery" of caveolae by Anderson's lab. Anderson and his colleagues have been studying the cytoplasmic surface of caveolae, which in electron micrographs shows up as a series of concentric ridges and furrows. More important, through antibody labeling of cells, they discovered that caveolae have a membrane marker protein, analogous to the clathrin that covers the coated pits, which they called caveolin. At the time, which was early last year, they even suggested caveolin might help isolate caveolae—a prediction that may already have been fulfilled.

Anderson's lab, however, was interested in more than caveolin. They concentrated much of their effort on a class of molecules known as GPI-anchored proteins, so named because they are anchored to the cell membrane by a lipid called glycosylphosphatidy inositol (GPI). Anderson's group found that GPI-anchored proteins, whose functions more and more biologists are growing curious about, cluster in caveolae. One such protein is the receptor for the B vitamin folic acid which is present in very low concentration in the blood stream. While studying caveolae in kidney epithelial cells, the Anderson's group discovered that after the folate receptors in caveolae bind the vitamin, caveolae close off their extracellular openings and can then dump a tiny flood of folate into the cytoplasm.

From this evidence, they proposed that caveolae are responsible for a new form of cellular uptake that they label potocytosis, with "poto" coming from a Latin verb that means "to drink." Unlike endocytosis, in which coated vesicles break free from the membrane and actually more into the cytoplasm, in potocytosis caveolae remain tethered to the membrane, closing their extracellular opening and thus creating a high concentration of ions that encourages their crossing the caveolae's membrane into the cytoplasm. Potocytosis appears to be the first clear-cut role for caveolae and Anderson contends many other small molecules and ions other than folate are delivered into the cell this way. Indeed, he suggests that concentrating and supplying dilute molecules to the cell is the central function of caveolae.

Not every cell biologist accepts that claim, however. Some are more than a little wary, contending that not even all epithelial cells make use of caveolae this way. "If potocytosis exists, it does not apply to endothelium," says Palade, who still maintains that transcytosis will ultimately prove to be one of caveolae's primary duties.

The apparent ability of caveolae to concentrate molecules that bind to receptors may be useful for more than just cellular uptake of metabolically necessary molecules; it also points to a role for caveolae as crucial sites where incoming signals from hormones, growth factors, and other extracellular regulatory molecules are brought into the cell. Michael Lisanti of the Whitehead Institute for Biomedical Research goes so far as to call caveolae "signaling organelles."

Lisanti and other investigators are now providing the experimental evidence to support this provocative notion. In recent experiments, for example, Kyoto University's Toyoshi Fujimoto and his colleagues in Japan labeled cells growing in culture with antibodies to identify proteins that concentrate in caveolae. They found enriched amounts of both a pump that removes calcium from the cell and a protein called the inositol trisphospate receptor, whose function may be to let calcium ions into cells. Since the caveolae contain machinery used in both calcium influx and outflux, Fujimoto has proposed that their role is to regulate calcium ion concentrations in the cytoplasm, a crucial task since calcium ions control many cell activities.

Other evidence implicating caveolae in signal transduction comes from an entirely unexpected source. In order to study cell structure, researchers for decades have washed cells with detergents such as Triton-100, which dissolves almost everything except the fibers of the cytoskeleton. This method, however, does leave behind insoluble protein-lipid complexes that can be separated from the cystoskeletal materials. Researchers have usually discarded these complexes as waste, but Whitehead's Lisanti now argues that they are isolated caveolae.

Lisanti began studying these complexes because they are rich in GPI-anchored proteins, his area of expertise. Earlier this year, he and his colleagues took electron micrographs of the insoluble material and got a surprise. The micrographs revealed spherical structures, 50 to 100 nanometers in diameters, that displayed surface features remarkably like those of the caveolae. That's when Anderson's identification of a marker protein like caveolin became crucial. Lisanti's group found that the concentration of that protein is 160 times higher in the complexes than in whole cells, clinching the case that they are actually caveolae, says Lisanti.

Through studying the molecules found in these detergent-insoluble complexes, Lisanti has become convinced caveolae play a major role in cell signaling. The complexes proved to be a rich source of proteins involved in various signaling pathways. In addition to those that had already been identified as being associated with the caveolae, Lisanti has found more than a dozen new ones. "We think caveolae are the region a lot of cell surface receptors will go to signal. I would bet more than half of the signaling molecules we know about will be there," he says.

Lisanti is already trying to trace at least one signal transduction pathway through the caveolae. He suggests caveolae may be the site where growth control signals are relayed to the Src family of tyrosine kinase enzymes. These enzymes add phosphate groups to other proteins, and in doing so, propagate growth control and other messages throughout the cell. Lisanti's research has shown that caveolin is a transmembrane protein, and he speculates that it may serve as the intermediary between GPI-anchored proteins and Src-like kinases. As Lisanti sees it, a ligand would bind to a GPI-anchored protein that in turn would transmit a signal to caveolin and from there to Src-like kinases.

That idea is not far-fatched. In the past 2 years, a number of investigators have obtained evidence indicating that many of these kinases do associate with GPI-anchored proteins. Moreover, the kinase produced by the src oncogene has been shown to phosphorylate caveolin. That might disrupt the normal processing of growth signals by caveolae, says Lisanti, and explain how oncogenic Src causes the unrestrained cell growth of cancer.

If Lisanti is right about the caveolae being centers for many signal transduction pathways, disruption of their function might well lead to other deleterious consequences. Some investigators caution, however, that caveolae's role in signal transduction is still in the realm of speculation. Much of it rests on the belief that the complexes Lisanti and others have isolated are in fact caveolae. But not everyone is convinced of that yet. "We're not automatically calling these detergent-insoluble complexes caveolae," says SUNY's Brown. She warns that washing cells and tissues with detergents like Triton is a harsh technique that may eliminate some key molecules that would normally be part of caveolae in the cell.

As investigators make further comparisons between proteins found in the complexes and those localized to caveolae by antibodies, it should become clear whether they have actually gotten their hands on isolated caveolae. Even assuming they have, cell biologists envision years of work before they understand the mechanics and role of these cellular vesicles. Most believe that caveolae will eventually be exposed as jacks-of-all-trades, serving multiple functions, perhaps within the same cell. "A lot is riding on these little pits," comments Doug Lublin of University of Washington. "This area is new enough that we're trying to define the structure at the same time we're trying to define its functions." It may be many years before biological spelunkers reveal all the riches buried in these little caves. ❑

****

Questions:

1. What are caveolae?

2. What is caveolin?

3. What is transcytosis?

Answers are at the end of the book.

Foreign invaders in the body are immediately confronted with a number of different defense mechanisms aimed at protecting you from disease. When an invader antigen strikes, it is attacked by a white blood cell called a B lymphocyte. Once inside the cell, it breaks down the foreign antigen protein and combines it with the body's own human-leukocyte-associated (HLA) antigen which is coded by the major histocompatibility complex (MHC), a group of genes with DNA sequences unique for each individual. Immunologists have discovered that the complex which forms, an organelle called CPL for 'compartment for peptide loading,' gets escorted to the membrane to be displayed on the surface of the cell as a signal for immune cells to attack. Scientists still have questions to ask such as what makes this new organelle and its contents so irresistible for attack.

Crucibles in the Cell

by Mark Caldwell

Back in Van Leeuwenhoek's day, when the microscope was a novelty, living cells defined the limits of little. It was inconceivable to biologists that they would one day intimately study structures *within* the cell, structures so small that beside them the cell looms like a swollen, lumbering giant.

Ah, such short sighted visionaries: these days biologists can glimpse tiny organelles under powerful electron microscopes. Of course, until recently that was just about *all* they could do: many cellular structures were simply too small to get hold of or get into and thus were difficult to characterize. Now new tools are changing that. Biologists can tell which organelles they're dealing with by sending out biochemical probes to detect the cell's constituent proteins. They can even show how the organelles work. And those workings often prove to have importance out of all proportion to the size of the worker.

One illustrative find this past year seems to have solved a major riddle of the immune system by revealing a key step in the way it detects foreign invaders. Last May researchers from the United States, the Netherlands, and Scotland announced that they'd identified a brand-new organelle—a submicroscopic crucible secreted deep inside one particular type of immune cell. Tiny though it is, the organelle appears to be responsible for assembling one of the immune system's most essential components: a tight little bundle of proteins that trigger the complex cascade of events that ultimately protect you from disease.

Immunologists have known for years that when a foreign invader, or antigen, first enters the body, it's swallowed by an immune-system sentinel: a white blood cell called a B lymphocyte. Once it has engulfed the antigen, the B cell minces it into bits of protein, or peptides, which it then brandishes on its surface as a signal to other immune cells. As the B cell then courses through the bloodstream, the immune system's attack cells, altered by the warning signals, spring into action. But there's a catch: the attack cells respond to a foreign peptide only of it's displayed in the clutches of a *nonforeign* protein called HLA in humans, and MHC Class II in the mice that immunologists usually study. These proteins are distinctively yours, marking you as yourself. Your immune system, after all, needs to know who's who; it doesn't want to attack everything indiscriminately, only foreigners that it perceives as having contaminated you.

What immunologists have *not* known, however, is just where in the B cell foreign antigens are broken down, and how the resulting peptides are combined with MHC to form a "complex"—a tight little bundle that squishes out onto the cell surface like a wad of bubble gum. Now it's clear that the hidden workshop is the newly discovered organelle of 1994. Thanks to their probes, the researchers were able to track both the antigen arriving in a cell and the MHC protein produced in the cell's nucleus and to see that the two paths merged at the hitherto unknown structure.

Cell biologist Sandra Schmid of the Scripps Research

Institute in La Jolla, California, has dubbed the organelle the CPL, for "compartment for peptide loading." She describes what goes on inside the CPL as a dynamic, multistep process, like cooking. "It's like a little pot of ingredients simmering on a stove," she says. Gradually self and nonself proteins are reduced and stirred into a new combination that's irresistible to the immune system's attack cells.

Of course, many questions about this process remain. But immunologists now know just where to search for the rest of the answers. ❑

****

Questions:

1. After being activated by antigen associated with a foreign invader, what do B lymphocytes do?

2. Why do the foreign antigen peptides combine with the nonforeign protein MHC?

3. What purpose does the organelle CPL serve?

Answers are at the end of the book.

The plasma membrane of the cell which forms the outer most boundary was first reported by cell biologists more than two decades ago. Its structure was proposed as a double layer of lipid molecules with proteins anchored on either side as well as some spanning the entire membrane surface. Other components such as cholesterol and carbohydrates could also be found within the plasma membrane. But when describing its appearance, scientists saw something quite interesting. The proteins were moving about the fluid membrane, changing its presentation like a mosaic pattern, thus it became known as the fluid mosaic model. This model has remained the framework for much of the work on membranes for many years. Current research on membranes using highly sophisticated techniques, however, is revealing new information which challenges some of the old ideas about membrane dynamics and the fluid mosaic model.

Revisiting the Fluid Mosaic Model of Membranes

by Ken Jacobson, Erin D. Sheets, Rudolf Simson

The fluid mosaic model, described over 20 years ago, characterized the cell membrane as "a two-dimensional oriented solution of integral proteins...in the viscous phospholipid bilayer" (*1*). This concept continues as the framework for thinking about the dynamic structure of biomembranes, but certain aspects now need revision. Most membrane proteins do not enjoy the continuous, unrestricted lateral diffusion characteristic of a random, two-dimensional fluid. Instead, proteins diffuse in a more complicated way that indicates considerable lateral heterogeneity in membrane structure, at least on a nanometer scale. Certain proteins are transiently confined to small domains in seemingly undifferentiated membrane regions. Another surprise is that a few membrane proteins undergo rapid, forward-directed transport toward the cell edge, perhaps propelled by cytoskeletal motors.

The more detailed view of the life of a membrane protein has emerged as a result of one old and two newer methods. For the past two decades, fluorescence recovery after photobleaching (FRAP) has been the major tool for measuring the lateral mobility of membrane components labeled directly with fluorophores or with fluorescent antibodies. In this method, a short pulse of intense laser light irreversibly destroys (photobleaches) the flourophores in a micrometer-sized spot. The flourescence gradually returns a fluorophores from the surrounding region diffuse into the irradiated area. FRAP experiments can reveal the fraction of labeled membrane proteins or lipids that can move, the rate of this movement (characterized by the lateral diffusion coefficient), and the fraction of proteins that cannot move on the time scale of the experiment. These apparently nondiffusing proteins are called the immobile fraction; a quantity that is frequently large and usually of unknown origin.

A second method, single-particle tracking (SPT), directly complements the information that is obtained from averaging the movement of hundreds to thousands of molecules in a FRAP experiment. In SPT, a membrane component is specifically labeled with an antibody-coated submicrometer colloidal gold or fluorescent particle, and the trajectory of the labeled molecule is followed with nanometer precision with digital imaging microscopy (*2, 3*). Visualization of individual protein motions can reveal submicroscopic membrane structures as the protein encounters obstacles in its path, although careful data analysis is required to distinguish between nonrandom and random movements (*4*).

The third method, recently applied to membranes, is the optical laser trap, allowing further characterization of the obstacles a membrane encounters. Proteins are labeled with submicrometer beads and manipulated in the plane of the membrane with laser light. Optical trapping occurs when a near-infrared laser beam with a bell-shaped intensity profile is focused on the bead attached to the protein. Optical forces on the bead, which are directed toward the highest intensity of the beam, trap the particle (*5*). By moving the laser beam or the microscope stage, the labeled protein can be dragged across the plasma membrane until it encounters a barrier or obstacle that causes the bead to escape the trap. The distance between barriers is called the barrier-free path (BFP).

The fluid mosaic model proposes random, two-dimensional diffusion for membrane components. Although lipids (*6*) and a fraction of the labeled protein population appear by SPT to diffuse freely, other protein movements are considerably more complicated than originally envisioned in the fluid mosaic model. One big surprise has been that a substantial fraction of the proteins are confined, at least transiently, to small domains. This has been seen most clearly for certain cell adhesion molecules [cadherins and neural cell adhesion molecules (NCAMs)] and nutrient and growth factor receptors. For cadherins, transferrin receptors, and epidermal growth factor receptors, the domains are 300 to 600 nm in diameter and confinement lasts from 3 to 30 s (*7*). Following earlier work on the red cell membrane, Kusumi and colleagues (*7*) proposed the "membrane-skeleton fence" model. In this scheme, a spectrin-like meshwork closely apposed to the cytoplasmic face of the membrane sterically confines membrane-spanning proteins to regions on the order of the cytoskeleton mesh size. Support for this model includes the facts that partial destruction of the cytoskeleton decreases the fraction of confined molecules, and truncation of the cytoplasmic domain leads to less confined diffusion (*8, 9*).

Can the fence model be supported by other techniques? Enter the laser trap. In a pioneering study, Edidin *et al.* (*10*) showed that BFPs for the lipid-linked and the membrane-spanning isoforms of the major histocompatibility antigens were ~1700 and ~600 nm at 23°C, respectively, and that these values increased with the temperature, indicating the dynamic nature of the barriers. Using weaker trapping forces, Sako and Kusumi (*11*) could detect even smaller BFPs for the transferrin receptor (~400 nm), which are consistent with the size of domains measured by SPT for both this receptor and E-cadherin (*7, 8*). The fences appear elastic, because the transferrin receptor rebounds after it strikes barriers (*11*), and a small fraction of these receptors seem to be fixed to the underlying cytoskeleton by spring-like tethers (*11*).

To permit the long-range diffusion observed by both SPT and FRAP, these barriers must open temporarily, either by dissociation of key molecular constituents of the barriers or by thermally driven local fluctuations of the meshwork-membrane distance. The escape of a given protein into an adjacent domain probably depends on the size of its cytoplasmic moiety, which implies that the effective domain size may be protein-dependent (*7, 12*). The emerging picture is that the immobile fraction of membrane proteins measured by FRAP does not simply represent stationary proteins but rather is some combination of proteins actually tethered to the cytoskeleton and those moving within and between confinement zones.

Is direct trapping by the cortical cytoskeleton the only means of confinement? Probably not. Surprisingly, confinement was also found for a lipid-linked isoform of NCAM in muscle cells, which cannot be directly trapped by the cytoskeletal network. In this case, the membrane domains were ~280 nm in diameter, and the proteins remained in them for about 8 s (*13*). The confinement may be the result of interactions with the same or other proteins that are associated with the cytoskeleton. SPT analysis suggests that the proteins in these zones are diffusing through a dense field of obstacles. Presumably, such domains will transiently trap different proteins, although this has not been proven. Other glycosylphosphatidyl inositol-anchored proteins such as Thy-1 also exhibit tightly confined diffusion (*14*), possibly because they are sequestered in glycolipid-enriched regions that include caveolae (*15*). Such confinement zones could play a significant role in mediating adhesion or in signal transduction by collecting relevant molecules, for example, cell adhesion molecules with their cooperating growth factor receptors (*16*).

A diverse set of membrane proteins can also be seen with SPT to move by highly directed, nondiffusional transport, sometimes in unexpected directions. Some proteins go in the direction opposite that of the bulk movement of patches of cross-linked proteins into caps seen in lymphocytes and other cells (*2, 17*). For example, integrins move outward toward the cell periphery in a highly directed fashion (*18*). These cellmatrix adhesion receptors, which are important for cell locomotion, may be recycled from the back to the front of the cell by forward-directed cytoskeletal motors.

The plasma membrane presents an intriguing mix of dynamic activities in which components may randomly diffuse, be confined transiently to small domains, or experience highly directed movements. The coexistence of multiple modes of diffusion and directed transport indicates pronounced lateral heterogeneity in the membrane. Key issues remain: How generally applicable is the membrane-skeleton fence model? Are transmembrane and glycosylphosphatidyl inositol-anchored proteins confined by the same structures in the cytoskeleton? How is the domain structure regulated? The greatest challenge will be to relate this exciting new knowledge of membrane dynamics to the manifold function accomplished by the plasma membrane. ❑

References and Notes

1. S. J. Singer and G. L. Nicolson, *Science* **175**, 720 (1972).
2. M. de Brabander *et al.*, *J. CELL BIOL.* **112**, 11 (1991); R. J. Cherry, G. N. Georgiou, I. E. G. Morrison, *Biochem. Soc. Trans.* **22**, 781 (1994) M. P. Sheetz, S. Turney, H. Qian, E. L. Elson, *Nature* **340**, 284 (1989).
3. R. N. Ghosh and W. W. Webb, *Biophys. J.* **66** 1301 (1994).
4. M. J. Saxton, *ibid.* **64**, 1766, (1993).
5. K. Svoboda and S. M. Block, *Annu. Rev. Biophys. Biomol. Struct.* **23**, 247 (1994).
6. G. M. Lee, F Zhang, A. Ishihara, C. L. McNeil, K. A. Jacobson, *J. Cell Biol.* **120**, 25 (1993).
7. A. Kusumi, Y. Sako, M. Yamamoto, *Biophys. J.* **65**, 2021 (1993).
8. Y. Sako and A. Kusumi, *J. Cell Biol.* **125**, 125 (1994).
9. R. N. Ghosh and W.W. Webb, *BioPhys. J.* **57** 286a (1990); Y. Sako, A. Nagafuchi, M. Takeich, A. Kusumi *Mol. Biol. Cell* **3**, 219a (1992).
10. M. Edidin, S. C. Kuo, M. P. Sheetz, *Science* **254** 1379 (1991).
11. Y. Sako and A. Kusumi, *J. Cell Biol.*, in press.
12. M. Edidin, M.C. Zuniga, M. P. Sheetz, *Proc. Nat. Acad. Sci. U.S.A.* **91**, 3378 (1994).
13. R. Simson *et al., Biophys. J.* **68**, A436 (1995).
14. E. D. Sheets, G. M. Lee, K. Jacobson, *ibid.*, p A306.
15. D. A. Brown and J. K. Rose *Cell* **68**, 533 (1992) R. G. W. Anderson, *Curr. Opin. Cell Biol.* **5**, 64 (1993).
16. E. J. Williams, J. Furness, F. S. Walsh, P. Doherty *Neuron* **13**, 583 (1994); C. O'Brien, *Science* **267** 1263 (1995).
17. B. F. Holifield and K. Jacobson, *J. Cell Sci.* **98** 191 (1991).
18. C. E. Schmidit, A. F. Horwitz, D. A. Lauffenburger, M. P. Sheetz, *J. Cell Biol.* **123**, 977 (1993).

19. We thank a number of colleagues for their helpful comments.

****

Questions:

1. What is the fluid mosaic model?
2. What groups of proteins have been found confined in small domains in the plasma membrane?
3. What does the "membrane-skeleton fence" model propose?

Answers are at the end of the book.

Homeostasis is essential for survival of cells, especially since they function only within a very narrow range of composition in the extracellular fluid. The buffer systems in the body contribute to homeostasis by maintaining the proper pH in the internal environment. The reaction of metabolically produced CO_2 with water to make bicarbonate and H^+ is an important reaction which is catalyzed by the enzyme carbonic anhydrase. Because the need to transport bicarbonate across the membrane is great, researchers have focused their efforts on studying the membrane-transport mechanism for bicarbonate. A new transporter is reported in this article which may be important for the control of intracellular pH regulation. Unlike the well characterized chloride/bicarbonate exchanger, this new system depends on the cotransport of potassium. Its exact physiological function, however, still remains to be determined.

Bicarbonate Briefly CO_2-Free

by Roger C. Thomas

Animal life is a remarkably complex mechanism for converting oxygen to carbon dioxide. The subsequent reaction of CO_2 with water to make bicarbonate and H^+ gives bicarbonate a central role in pH homeostasis. Numerous different membrane-transport mechanisms for bicarbonate are known, and another one emerges from the ingenious studies with out-of-equilibrium solutions reported by Zhao and colleagues.

The slowness of the reaction of CO_2 with water is overcome in physiology by the enzyme carbonic anhydrase.[2,3] This was famously first found in red blood cells where it maximizes the capacity of blood to carry CO_2 (as bicarbonate). The first bicarbonate transporter recognized was indeed the chloride/bicarbonate exchanger of the same cell. It allows the rapid interchange of bicarbonate and chloride, to generate the chloride shift. Since then many other bicarbonate transporters have been discovered, the one of most general importance in intracellular pH regulation probably being the Na-dependent Cl^{-1} HCO_3^- exchanger.[4,5]

A similar mechanism regulates pH in squid giant axons[6] but bicarbonate is still carried across the axon membrane in the complete absence of Na^+ or Cl^- ions.[1] It proved difficult to investigate the mechanism involved using conventional CO_2/HCO_3^- solutions, because the CO_2 makes it impossible to maintain a significant bicarbonate gradient. The normal bicarbonate saline of pH 8 used in squid-axon experiments contains 12 mM bicarbonate and must be equilibrated with 0.5% CO_2 to be stable. If this equilibrium is disturbed, for example by the loss of CO_2 to the atmosphere, the pH will slowly increase as bicarbonate too is converted to CO_2 and lost.

A stable bicarbonate solution must therefore contain CO_2. This crosses cell membranes so fast that its level inside cells is only rarely different from that outside. The intracellular CO_2 is in equilibrium with intracellular H^+ and HCO_3^-, often catalyzed by the almost ubiquitous carbonic anhydrase. So you cannot normally have external bicarbonate without a similar level inside cells. The exact level will depend on the intracellular pH, but this is usually close to the external pH.

Zhao and colleagues[1] solved the problem by the clever use of out-of-equilibrium solutions. They mixed two pairs of carefully formulated solutions in a continuous-flow apparatus which delivered the mixture to the outside of a squid axon within half a second. They mixed either a 24 mM bicarbonate alkaline solution or a 1% CO_2 acid solution with acid or alkaline CO_2/bicarbonate-free solutions to give pH 8 mixtures with 12 mM bicarbonate or 0.5% CO_2. The rate constants of the conversion of bicarbonate to CO_2 and vice versa are such that very little takes place within half a second. All solutions contained acetazolamide to inhibit any carbonic anhydrase that might be present.

That the CO_2-free, freshly mixed solution indeed contained almost no CO_2 is shown convincingly by its lack of effect on the intracellular buffering power, which Zhao *et al.* measured by applying ammonium solutions. The equilibrated CO_2/HCO_3^- solution increased buffering power by 57%, as expected, whereas the out-of-equilibrium solution increased it by only 7%.

In contrast to its lack of effect on buffering, the CO_2-free, out-of-equilibrium bicarbonate solution doubled bicarbonate uptake rates by potassium-depleted axons in high-K solution, strong evidence for cotransport of K and HCO_3^-. Similarly, bicarbonate efflux from K-loaded axons in bicarbonate-free solutions was much faster if the bathing solution contained CO_2. The CO_2 outside it converted to about 2 mM bicarbonate inside, so there is a large outward bicarbonate gradient as well as K gradient to drive the cotransporter.

The physiological function of this hitherto unknown K/HCO_3^- cotransporter is obscure. But the new technique applied by Zhao *et al.* may breathe fresh life into the study of bicarbonate transport generally. ❑

1. Zhao, J., Hogan, E. M., Bevensee, M. O. & Boron, W. F. *Nature* **374**, 636-639 (1995).
2. Maren, T. H. *Physiol. Rev.* **47**, 595-781 (1967).
3. Widdas, W. F. Baker, G. F. & Baker, P. *Cytobios* **80**, 7-24 (1994).
4. Thomas, R. C. *J. Physiol., Lond.* **273**, 317-338 (1977).
5. Schwiening, C. J. & Boron, W. F. J. *Physiol., Lond.* **475**, 59-67 (1994).
6. Boron, W. F. & Russell, J. M. *J. Gen. Physiol.* **81**, 373-399 (1983).

****

Questions:

1. What reaction plays a central role in pH homeostasis in the cell?
2. What enzyme catalyzes the reaction of CO_2 with water?
3. What ion is necessary to drive the cotransporter for bicarbonate in this system?

Answers are at the end of the book.

Part Two:

What's New in Tissue Engineering

Spinal cord injuries are tragic because they can cause permanent neurological damage. But one mystery that needs to be solved is why these central neurons fail to regrow like the peripheral nerves do after being cut. Scientists are focusing their attention on a protein called myelin-associated glycoprotein (MAG), found in the fatty myelin sheath that surrounds nerve fibers, as a major inhibitor of neurite growth. Curiously enough, MAG is also present in the peripheral nervous system but in lesser quantities than found in the central nervous system. By using animals that lack a functional MAG gene, neuroscientists will be asking definitely if central nerves can be regenerated. Finding the missing piece to this puzzle will most certainly lead to better strategies for treating spinal cord injuries in the future.

Old Protein Provides New Clue To Nerve Regeneration Puzzle

by Marcia Barinaga

If you cut a nerve in your arm or leg, all is not lost. The cut end of the nerve will send out new shoots called neurites, which grow along the path of the old nerve and restore some neuronal function to the limb. But a cut spinal cord is another matter entirely. While it, too, will sprout tiny neurites, they won't grow across the damaged area. Therein lies a tragic puzzle: Why will peripheral nerves regenerate, while those of the central nervous system (the brain and spinal cord) remain severed, leaving an injured person with permanent neurological damage? Finding the answer to that question promises big rewards, as it may lead to better strategies for treating spinal cord injuries.

Now, two research teams, one including Marie Filbin of the City University of New York and Patrick Doherty and Frank Walsh of Guy's Hospital in London, and the other including Lisa McKerracher, Sam David, Peter Braun, and their colleagues at McGill University in Montreal, have made an advance towards solving this puzzle. The groups independently discovered that a protein known as myelin-associated glycoprotein (MAG), found in the fatty myelin sheath that surrounds nerve fibers, blocks neurite growth in cell culture and may therefore be at least partly responsible for the failure of central neurons to regrow when damaged. The Filbin team reports it results today in the September issue of *Neuron*; the McGill researchers will describe theirs in *Neuron*'s October issue.

The discovery came as a surprise because MAG, which was discovered in the 1970s, had been shown to stimulate rather than inhibit the growth of young neurons—just the opposite of what the new work finds with adult neurons. Although these results in the culture dish have yet to be confirmed in living animals, researchers are hoping that the finding will place them a bit closer to a strategy for mending damaged spinal cords. "The clinical significance is still unclear, but it is definitely a step forward, and very exciting," says developmental neurobiologist Marc Tessier-Lavigne of the Howard Hughes Medical Institute at the University of California, San Francisco.

The search for molecules that block regeneration dates back to the early 1980s, when Albert Aguayo and his colleagues at McGill University found that, although central nervous system (CNS) neurons don't regenerate in nature, they do have the intrinsic ability to do so, if conditions are right. When the researchers transplanted pieces of peripheral nerve across damaged regions of brain and spinal cord in rats, some of the damaged neurons sent neurites across the transplanted "bridge." Only a small percentage of nerve fibers made the trek—but that was better than what the neurons would have achieved on their own.

If CNS neurons are able to regenerate, then something must be stopping them from doing so under normal conditions. At least part of the inhibition

Reprinted with permission from *Science*, September 23, 1994, Vol. 265, pp. 1800-1801.

seems to come from the myelin that covers CNS nerves. In 1988, Martin Schwab of the University of Zurich found that CNS myelin inhibits neurite growth from a variety of neurons grown in culture. He went on to partially purify an inhibitory protein he calls IN-1, although it couldn't account for all of the effect. When he used antibodies to block IN-1 activity in rats, severed spinal neurons sent out neurites, but only 5% to 10% of the cut neurons crossed the gap. This, says Schwab, may mean that "other [inhibitory] molecules are there."

The Canadian group began searching for those other molecules about 2 years ago. McKerracher and David, who had experience in neuron culture, joined forces with Braun, a biochemist with experience in purifying proteins from myelin. They found two fractions of myelin protein that blocked neurite growth in an immortalized neuronal cell line growth in culture.

In one of these fractions the inhibition seemed to be due to MAG; anti-MAG anti-bodies, which bind to protein, could completely strip that fraction of all of its inhibitory activity. That was difficult to believe at first, both because of MAG's known ability to stimulate neuronal growth and because it is found in the myelin of both peripheral (PNS) and CNS neurons, which would seem to rule it out as a CNS-specific inhibitor. "People have been suspecting for a while that there will be more than one inhibitory molecule," says Roger Keynes, who studies regeneration and neural development at the University of Cambridge, England. But, he adds, "certainly people weren't thinking about [MAG]....It was definitely a surprise."

Further experiments showed that the surprising result was in fact correct. When the Canadian group tested a pure fragment of MAG protein provided by Robert Dunn, also of McGill, they found that it too blocked neurite growth. Finally, the team used anti-bodies to remove all the MAG from a preparation of myelin protein and found that neurite growth on the remaining proteins was resorted to 63% of what it was in the absence of any myelin protein. They concluded that MAG is a major inhibitor of neurite growth.

Meanwhile, Filbin and her collaborators reached the same conclusion—independently and from a very different direction. Their initial plan was to study the stimulatory effects of MAG that others had observed with young, developing neurons, such as sensory neurons from the dorsal root ganglia (DRG) in the spinal cords of newborn rats. To mimic the way MAG is produced by the cells that make myelin, Filbin first engineered hamster cells to make the protein and then took the cell line to the London labs of Doherty and Walsh, who are experienced in studying neurite growth. The team tested the effects of the MAG-making cells of neurons from the cerebellum of newborn rats, because they happened to have those cells handy at the time. They were shocked to find that cerebellar neurons would not grow over the MAG-producing cells, although they would grow happily on control cells. "We didn't believe it," says Filbin. But the result held up.

The team then repeated the experiment with DRG neurons from newborn rats, as others had done, and found that MAG encouraged growth of those neurons, as expected. But they realized that neuron age can make a difference; Jim Cohen at Guy's Hospital had recently found that adult DRG neurons will not grow on cultured strips of tissue from peripheral nerves, although young DRG neurons will. When Filbin and colleagues tested older DRG neurons from 5- to 7-day-old rats, the neurons refused to grow on the MAG-producing cells.

"The two papers are beautifully complementary," says Tessier-Lavigne. "Together they each answer the major questions raised by the other's work." The Montreal group studied MAG's effects on an immortalized neuronal cell line, whose properties could differ from true neurons. Filbin, Walsh, and Doherty found, however, that the protein has the same effects on neurons that have been recently removed from animals. The Filbin group's experiments could not determine the degree to which MAG contributes to myelin's overall inhibitory effect, an answer that was filled in by the Montreal team, when they showed that MAG accounts for well over half of myelin's inhibitory activity.

But that leaves a third major question: As MAG is present in PNS myelin, why are peripheral nerves able to regenerate? The answer may lie partly in the fact that PNS myelin contains one tenth as much MAG as does CNS myelin. In addition, MAG is probably not around when peripheral nerves regrow. That's because damaged peripheral nerves undergo a rapid cleanup that removes all of the myelin debris downstream of the nerve cut. Only after the debris is gone can new neurites grow into the injured area, says Filbin. Cleanup of CNS neurons, on the other hand, is much slower, so MAG-containing myelin stays around much longer after nerve injury.

Still, MAG accounts for only 63% of the neurite inhibition from myelin, although the Montreal group has another, non-MAG-containing protein fraction that also has inhibitory effects. The culprit in that fraction could be Schwab's IN-1, McKerrachers says.

If MAG ia part of the reason why CNS neurons won't regenerate, then tricks for blocking MAG function may someday be useful for treating spinal cord injuries. But before researchers start thinking along those lines, they first must show that blocking MAG facilities neuronal regrowth in animals with CNS injuries. The Montreal group has begun such experiments with John Roder of Mount Sinai Hospital in Toronto.

Roder's lab recently made mutant mice that lack a functional MAG gene, and the group plans to see whether CNS nerves can be coaxed to regenerate in the MAG-minus mice. "That's the acid test," says Roder. If MAG passes it, then researchers on the long trek toward nerve regeneration will have taken a significant step forward. ❑

****

Questions:

1. What protein inhibits neurite growth in the central nervous system?

2. Where is MAG found in the body?

3. Why are peripheral nerves able to regenerate in the presence of MAG?

Answers are at the end of the book.

Neurobiologists are very excited about neurotrophic factors, naturally occurring proteins that help keep nerve cells alive and healthy. The news sounds good to the many people who suffer debilitating neurodegenerative diseases, including Alzheimer's, Parkinson's and amyotrophic lateral sclerosis (ALS), or Lou Gehrig's disease. How do the neurotrophic factors fit into the scheme of things as therapeutic agents? One example showing benefits is the administration of nerve growth factor (NGF) to rats whose cholinergic neurons are damaged like that of Alzheimer's patients. Although rats are not the perfect animal models for Alzheimer's, their neurons could be saved from dying by infusions of the nerve-nurturing protein NGF. Factors like NGF, insulin-like growth factor I and ciliary neurotrophic factor, may also be beneficial in treating drug-induced peripheral neuropathies so commonly experienced in diabetes and cancer chemotherapy. The sooner researchers determine the most effective way to deliver these natural proteins to the body, the better will be the survival of many neurons throughout the body and brain.

Neurotrophic Factors Enter the Clinic

The biotech industry launches a new class of nerve-nurturing drugs with high hopes of toppling stubborn neurological diseases such as Lou Gehrig's disease

by Marcia Barinaga

Neurons are among our most precious cells. They play vital roles in our lives, governing everything from the recoil of a finger from a hot stove to the understanding of Dante. Unlike the vast majority of other cells, however, most neurons must perform these all-important tasks for an entire lifetime, as they aren't replaced if destroyed by injury or disease. Given that neurons are so precious, it's not surprising that one of the hottest areas in neuroscience today is the study of neurotrophic factors: naturally occurring proteins that keep neurons alive and healthy during embryonic development and later in normal adult life.

This interest in neurotrophic factors is not merely academic. In fact, it has already generated a "wave of excitement" in the biotech world, says Jeff Vaught, senior vice president for research at Cephalon, a biotech company in West Chester, Pennsylvania. A half-dozen or more companies have clinical trials planned or under way for testing neurotrophic factors against debilitating neurodegenerative diseases, including Alzheimer's, Parkinson's, and amyotrophic lateral sclerosis (ALS), or Lou Gehrig's disease.

"Most of the diseases for which these neurotrophic factors are being developed are diseases for which current therapies are virtually nonexistent," says biotech stock analyst Margaret McGeorge of Hancock Institute Equity Services in San Francisco. That lack of treatments, she says, is a driving force behind the current enthusiasm. But McGeorge also points out that "it is much too early to say" which factors will pan out as useful drugs. One highly touted neurotrophic factor has already stumbled in clinical trials — possibly a result of being rushed into large-scale trials. That experience hints that bringing these nerve-nurturing drugs to market could pose bigger challenges than companies originally thought. Still, that setback has only slightly dimmed the excitement that surrounds neurotrophic factors in the research and biotech communities.

That excitement has been building to its current crescendo for a long time, beginning more than 30 years ago with the Nobel prize-winning discovery of the first neurotrophic factor — nerve growth factor (NGF) — by Rita Levi-Montalcini, Stanley Cohen, and Viktor Hamburger at Washington University in St. Louis. They showed that NGF ensures the survival of certain peripheral neurons as they grow toward and connect with target tissues during the development of the nervous system. Later work showed that NGF is needed for the survival of some brain neurons as well.

Reprinted with permission from *Science,* May 6, 1994, Vol. 264, pp. 772-774.

But many types of nerve cells didn't respond to NGF, raising the possibility that they are supported by other factors. The search for those factors has so far uncovered roughly a dozen proteins that help neurons survive throughout the body and brain. Some, like NGF, work only on some types of nerve cells. Others, such as insulin-like growth factor (IGF) and fibroblast growth factor (FGF), were originally identified as growth factors for other tissues, such as muscle, or the immune system, and were only later found to nurture neurons as well. As the field has grown, the definition of a neurotrophic factor "has been expanded a lot," says Washington University neuroscientist Eugene Johnson, "and it's clear that there are a lot more players."

ALS: A cautionary tale

As this cast of characters expanded, researchers quickly realized that some of the newly discovered growth factors might serve as therapeutic agents for neurological diseases. The first disease to be targeted by neurotrophic factors in a major clinical trial was ALS, a deadly condition in which the motor neurons — neurons that control the skeletal muscles — progressively degenerate, eventually depriving patients of all movements, even the ability to breathe.

ALS was a good starting point for several reasons. For one thing, there's no effective treatment for the disorder. In addition, motor neurons, unlike neurons of the brain, are accessible to protein drugs introduced into the bloodstream. Says Cephalon's Vaught, diseases like ALS should be "very amenable to growth-factor application because the projections from the affected neurons lie outside the blood-brain barrier," the physiological wall that keeps large molecules like proteins from entering the brain.

A number of growth factors had already been shown to support the growth of motor neurons in laboratory cultures, and three of those factors became early candidates for treating ALS: brain-derived neurotrophic factor (BDNF) and insulin-like growth factor I (IGF-I), both made by muscles controlled by motor neurons, and ciliary neurotrophic factor (CNTF), made by the Schwann cells that form an insulating sheath around the neurons.

In addition to the cell-culture work, all three factors also turned out to aid the healing of injured motor neurons in rats. Moreover, the factors improved the condition of mice with hereditary motor-neuron diseases similar to ALS. That was enough for two companies, Regeneron Pharmaceuticals of Tarrytown, New York, and Synergen of Boulder, Colorado, to begin clinical trails of CNTF and for a third, Cephalon, to begin trials with IGF-I. Regeneron is also developing BDNF, in collaboration with Amgen of Thousand Oaks, California, although they have not yet begun large-scale trials.

But the high hopes with which these trials began quickly turned into what many observers view as a cautionary lesson for the field. The first publicly reported results with CNTF looked promising enough: Last September, at the World Congress of Neurology in Vancouver, British Columbia, Regeneron announced that 12 ALS patients who had been receiving CNTF injections in a small, preliminary trial showed less of a decline in muscle strength than did 14 patients receiving a placebo. The results were not statistically significant, Jesse Cedarbaum, vice president for clinical affairs at Regeneron, told *Science* in an interview last November, but they were sufficiently encouraging to justify a full-scale trial of CNTF's effectiveness. Last year Regeneron began that trial, which includes 720 patients at 36 clinical sites.

But even as Regeneron trumpeted its early results, optimism about CNTF had already begun to erode. Rumors were circulating that patients in a preliminary safety trial of CNTF carried out by Synergen had suffered side effects, including coughing, fever, weight loss, and activation of herpes virus, which can hide in latent form in some neurons.

Similar side effects were also rumored to be plaguing Regeneron's large-scale trial — rumors that were soon to be borne out. In March, Regeneron researchers announced that they had unblinded the data on 550 patients who had completed 6 months of the trial and found that a substantial number of those receiving CNTF had not only had serious side effects, but had actually fared worse on measures of muscle strength than did patients receiving placebos. Patients who didn't experience side effects showed modest improvement as compared to controls, but investors weren't mollified by that scrap of good news, and Regeneron's stock dropped 50%. The company has tried to salvage the trial by applying to the Food and Drug Administration to continue it with lower doses aimed at reducing the side effects.

But the question remains whether anything more than modest effects can be expected of CNTF. Synergen has a large trial in progress, using lower drug doses than Regeneron, based on the findings of its safety trials. The trial has not been unblinded, but insiders say the prospects don't look very exciting: "Although all the numbers haven't been added up [for either large-scale trial], the clinical results in ALS are less than we had hoped for," says Tufts University neurologist Theodore Munsat, who has participated in the Synergen trials. The best that can be expected, says Munsat, is "a modest slowing of the deterioration rate." He questions whether such modest effects would improve the quality of a patient's life enough to justify use of the drug.

But some researchers think the first round of trials is not a fair test of CNTF. Neurologist Michael Sendtner of the Max Planck Institute for Psychiatry in Martinsried, Germany, argues that the way the drug is administered could be key to reducing its side effects and improving its effectiveness. He recently found, for example, that CNTF has a half-life of only 3 minutes in the bloodstream of a rat. That means the injected drug has little time to reach the motor neurons. While in the bloodstream, it encounters tissues such as the liver and lungs and also interacts with blood-borne immune cells. CNTF's interaction with these cells and tissues, which don't usually encounter the factor, may account for its side effects, says Sendtner. He suggests there is still hope for better results if CNTF were delivered directly to the cerebrospinal fluid, which bathes the spinal roots of the motor neurons.

Even if CNTF doesn't pan out, biotech researchers hope IGF-I will have fewer drawbacks as an ALS therapy. That drug has already been used in large-scale drug trials for conditions including diabetes, dwarfism, and osteoporosis without showing serious side effects. "The good news about IGF-I is that it certainly appears so far to be fairly benign," says biotech analyst Timothy Wilson of the New York securities firm Hambrecht & Quist. "But the big question mark is whether or not it will have efficacy in ALS." The Cephalon trial aims to answer that question.

But even if other companies do hit pay dirt with an ALS treatment, close observers of the field think Regeneron's experience holds some lessons for all the companies involved. Some argue, for example, that Regeneron moved precipitously, relying on results from early trials that were too small for adequate analysis of optimal dosing. "When companies have tried to move too quickly through these clinical trails, they usually find disappointment," says R. Brandon Fradd, a biotech stock analyst with Montgomery Securities in San Francisco. But Fradd and others emphasize that, despite the first disappointing results, the field of neurotrophic factors still shows promise.

Peripheral neuropathy: Better odds

Some of that promise may be realized in tackling a more straightforward problem. One reason finding ALS therapies is so difficult is that the cause of the motor neuron degeneration remains unknown, so there is no good animal model for the disease. The story is different for peripheral neuropathies: the deterioration of sensory or motor neurons caused by diabetes or by treatment with cancer chemotherapeutic agents such as vincristine, cisplatin, and taxol. For those conditions, there are good models. "You can cause peripheral neuropathy with vincristine in a rat in man," says Wilson, "and if IGF-I prevents it in the rat, there is a very good chance it will do so in man, because it's the same chemical insult."

And this condition has already been met, since IGF-I does prevent drug-induced peripheral neuropathies in animals, as does NGF. Jack Kessler, Stuart Apfel, and their colleagues at Albert Einstein College of Medicine in New York have collaborated with Cephalon and with Genentech of South San Francisco to show the effectiveness of these two factors against drug-induced neuropathies in mice.

As in the rest of this field, clinical trials are following very quickly behind laboratory results; Cephalon is planning to begin clinical trials of IGF-I for peripheral neuropathy within a few months. Genentech has already completed a small multicenter dosing and safety trial of NGF in human patients with peripheral sensory neuropathy, and expects to begin a larger trial by this summer, which would include diabetics with sensory neuropathy. They are tackling diabetes first, says Kessler, who is participating in the studies, because "diabetic neuropathy is a much more prevalent and disabling medical problem" than neuropathy from cancer chemotherapy. As a result of neuropathy, diabetics often experience ulceration of their feet and legs, which can eventually require amputation.

But those two neurotrophic factors aren't the only ones being tried out against neuropathy. Genentech scientist Gao Wei-Qiang recently demonstrated that a factor called neurotrophic-3 (NT-3) protects mice from neuropathy caused by the anti-cancer drug cisplatin, says Franz Hefti, director of Genentech's neuroscience department. The upshot, says Hefti, is that "for peripheral neuropathy a combination of neurotrophins is likely to do the trick."

Parkinson's and Alzheimer's: A fresh approach

Most of the clinical trials now underway have concentrated on testing neurotrophic factors as treatments for conditions such as ALS and peripheral neuropathy in which peripheral nerve cells deteriorate. But that kind of therapy doesn't exhaust the possible uses of the trophic factors. One other important area where these proteins are now beginning to play a role is as experimental therapies for neurodegenerative diseases of the brain.

Alzheimer's is the most prevalent of those diseases, and the leading candidate for Alzheimer's therapy is NGF. An early clue that NGF might be of benefit came from experiments on one of the major groups of neurons that degenerate in Alzheimer's — the cholinergic neurons. These neurons (so called because they release acetylcholine as a neurotransmitter) send their projections into a memory center in a brain structure called the hippocampus. Several research groups have shown that when cholinergic neurons in rat brains are damaged by having their projections cut, they can be saved from dying by infusions of NGF.

Further experiments suggest that saving those neurons may have consequences for mental function. When Fred Gage and coworkers at the University of California, San Diego, infused NGF into the brains of aging rats that showed memory impairment, they found what Gage describes as "very good improvement in [learning] behavior." On a test of memory retention, he adds, the NGF-treated rats "were in the range of the aged nonimpaired group."

Encouraging as these results are, it's far too soon to imagine that NGF will do the same thing for Alzheimer's patients that it did for aging rats. For one thing, the rats are not a perfect animal model for Alzheimer's. Furthermore, it isn't clear whether the decay of the cholinergic neurons is in fact a cause or an effect of human Alzheimer's. Nevertheless, the animal results suggest that keeping the cholinergic neurons alive can improve mental function, says Genentech's Hefti.

Given that possibility, Alzheimer's researchers are ready to see what effects the neurotrophic factor might have in human beings. Lars Olson and his colleagues at the Karolinska Institute in Stockholm have already tried NGF infusion in one Alzheimer's patient; they report that the patient showed improvement on a memory test. Both Genentech and Synergen, the latter in collaboration with Syntex, are planning clinical trials of NGF for Alzheimer's disease as soon as next year. "We don't expect NGF to cure the disease," says Hefti, "but we expect NGF to have a significant behavioral effect."

Researchers also have high hopes for neurotrophic factors in the treatment of Parkinson's disease, another common and intractable neurodegenerative disease, which is characterized by the degeneration of certain dopamine-producing neurons in the brain. Last year, Frank Collins and his colleagues at Synergen purified a protein they call glial cell-line derived neurotrophic factor (GDNF), which supports the survival, in cell culture, of the neurons that die in Parkinson's disease. Using GDNF, "we are hoping that we can reverse the disease" says Synergen chief executive officer Larry Soll. GDNF won't bring a dead neuron back to life, he notes, but it might do something almost as good: Experiments in rats and mice by Synergen researchers suggest GDNF can coax surviving neurons to send out new projections to replace those from lost neurons. If it has similar effects in primates, Soll says Synergen hopes to have GDNF in clinical trials by 1995.

Although the future of any single therapeutic approach with neurotrophic factors remains uncertain, the possibilities are numerous enough to fuel several generations of clinical trials, using increasingly sophisticated delivery systems. And even if the first round comes up short of expectations, observers say it is only a matter of time until our besieged neurons find new pharmaceutical allies in the form of proteins designed by nature to keep neurons alive. ❑

Additional Reading

S. Korsching, "The Neurotrophic Factor Concept: A Reexamination," *Journal of Neuroscience* **13**, 2739 (1993).

H. Thoenen, "The Changing Scene of Neurotrophic Factors," *Trends in Neuroscience* **14**, 165 (1991).

P. H. Patterson, "The Emerging Neuropoïetic Cytokine Family: First CDF/LIF, CNTF and IL-6; Next ONC, MGF, GCSF?" *Current Opinion in Neurobiology* **2,** 94 (1992).

****

Questions:

1. What are neurotrophic factors?
2. How can neurotrophic factors be used to treat neurodegenerative diseases?
3. Why is the route of neurotrophic factor administration different for different diseases?

Answers are at the end of the book.

For the millions of people who suffer organ failure and tissue loss there is a tremendous need for healthy, transplantable tissue. Tissue engineering may sound like something out of a science fiction novel but it has become the vision of many researchers who believe that one day we will be able to replace damaged organs with new tissues. Live cells, grown and supported in the laboratory, can be molded into new ears, noses and ribs for animals and will soon be tested in humans for its applicability. Engineered bones and tendons are also being tested for their potential benefits. Most notably is the use of engineered dermal cells, the equivalent of living gauze, which can cover open wounds and stimulate the body's own healing. Further down the road is the prospect that kidney, heart and liver functions could be restored by injecting cells into critical areas where they are needed. Still in the infancy stages, tissue engineering is likely to become the next best thing to genetic engineering.

Tissue Engineering

Replacing damaged organs with new tissues

by Richard Lipkin

In a housing project on the outskirts of Chicago, a three-alarm fire erupts. Engines rush in, and firefighters valiantly pull a woman and her 6-year-old twin sons from the burning building. One of the boys has third-degree burns over a large part of his body. His condition is critical. Without a supply of fresh skin to cover his wounds, he is likely to die.

Meanwhile, at home in Canton, Ohio, a retired veteran, his ailing liver about to give out, waits anxiously for the phone to ring. His name ranks 127th on a list of potential recipients of donor organs, a list that moves far too slowly for his eroding condition. Without a liver transplant, he may not survive the month.

In a townhouse community in Irvine, Calif., a 38-year-old mother moves groceries into the refrigerator. Suffering from a rare form of arthritis, she finds the mere act of putting away a quart of milk excruciating. The cartilage in her joints has worn away prematurely, and her medications do little to alleviate the intense pain. The solution, her physicians advise, lies in the ability of surgeons to replace tissue in key weight-bearing joints.

Though these medical cases may seem to have little in common, all three people suffer from the same general problem Their bodies cannot repair their injured organs. Moreover, their conditions are curable. What they need is healthy, transplantable tissue.

Enter the tissue engineers. These scientists hail from a wide range of disciplines — materials science, chemical engineering, molecular biology, and specialized surgery, to name only a few. Together, they have formed diverse research teams with the aim of learning how to grow new tissue in the laboratory to replace damaged tissue unable to repair itself.

Regeneration of damaged body parts, once considered the province of science fiction, is now moving toward science reality, says Robert Langer, a chemical engineer at the Massachusetts Institute of Technology. Though physicians can by no means regenerate a complete new heart, liver, or kidney, they have grown parts of these and other organs.

For less complicated tissues, such as skin, cartilage, and tendon, several research teams have made substantial headway. They have succeeded, in a number of cases, in growing simple tissues outside the body and then transplanting them into animals. For a few such tissue replacements, clinical trials with human patients are poised to begin.

"There's a tremendous need out there for transplantable tissues and organs," says Gail K. Naughton, chief scientist for Advanced Tissue Sciences in La Jolla, Calif., "When you see the kids, the ones with severe burns, you just want to do something. They keep you going during the long hours.

"There's a large gap between the number of people who need transplants and the number of available organs," she adds. In addition, technical problems and legal threats have caused manufacturers to withdraw from the market some implantable devices, such as artificial joints and breast implants.

"People are only beginning to realize the potential here," says Naughton. "Tissue engineering today is where genetic engineering was 10 years ago."

Organ failure and tissue loss affect millions of people annually, causing premature deaths and immense suffering. Each year in the United States, physicians perform an estimated 8 million surgical procedures to treat these conditions, which require more than 40 million days of hospitalization and cost an estimated $400 billion. Usually, the number of people needing organ transplants exceeds the number of donated organs, sometimes by a factor of 10. In 1989, for instance, liver disease led to 30,000 deaths, though only 2,160 transplants were performed — a shortage brought to public attention by the recent liver transplant of 63-year-old baseball legend Mickey Mantle.

People in need of other organs often fare no better. Many of the 600,000 people with kidney ailments, the 728,000 with pancreas disorders, and the 150,000 with severe burns await new organs. Patients requiring tissue for rarer procedures, such as the 30,000 persons annually who need cartilage for facial reconstructive surgery or the 33,000 in need of tendon repair, find themselves in a similar bind.

"From our point of view, the end goal is to create new tissues to serve as permanent replacements," says Joseph P. Vacanti, a pediatric surgeon at Harvard Medical School and Children's Hospital in Boston. "Ultimately, we want to grow tissue from a patient's own cells to avoid an immune reaction. But to achieve that goal, there's a lot of work to be done."

A major obstacle, says Vacanti, is determining exactly how and why cells differentiate. Most organs include many types of cells, each with its own distinct characteristics. Together, those cells interact harmoniously, producing a whole, functioning organ.

"How do cells know how to reorganize into a new structure?" Vacanti asks. "How do blood vessels know how to grow into a new tissue? How does an organ know when to stop growing? Those are basic scientific questions that need to be answered for any tissue under study."

If researchers can discern general scientific principles from a few tissue regeneration models in the laboratory, perhaps those principles could be applied to other organs in the body, Vacanti points out.

"If the experiments continue to look promising, I expect that nearly every human tissue will, at some point, be amenable to tissue engineering," he says.

"That's my vision of the future."

To handle certain urinary tract problems, Anthony Atala, a surgeon at Harvard and Children's Hospital, has organized a research team to grow replacement tissues for the ureter, bladder, urethra, and kidney. For example, to treat the congenital condition of ureteral reflux, which causes urine to back up into the kidney rather than flow directly into the bladder, the scientists are growing tissue to augment weak ureters and halt the ensuring kidney damage. "Thirty years ago, reflux was the main cause of kidney failure," Atala says. "Now we are treating it with engineered cells."

To strengthen or enlarge weak bladders, Atala's team has developed a method of injecting muscle cells into the walls of a bladder. "Instead of having painful open surgery and a week-long hospital say, patients can be treated with a relatively simple procedure in a physician's office." To enlarge bladders, the researchers are growing strips of artificial tissue that surgeons can stitch into the organ's wall.

Some of these treatments are ready for testing on humans, says Atala, pending Food and Drug Administration approval. Looking further ahead, scientists are attempting to create artificial kidney tissues. "The kidney is more difficult because it has many tissues with different functions," Atala says. "But if we could replace even one or two of a damaged kidney's missing functions, that would be an improvement."

Atala thinks that the basic idea of injecting engineered cells could be used to reconstruct other organs damaged by disease or by surgery for cancer. Several groups in collaboration with Reprogenesis in Dallas are at work on methods of reconstructing breast tissue after mastectomy, he adds. "There's a strong need for this kind of procedure, especially given the problems surrounding breast implants."

Many of the tissue engineering techniques have emerged from a relatively simple concept that originated with Langer and Vacanti more than 10 years ago. The idea is to make gossamer sheets of dissolvable material, seed them with cells, then incubate them in a bioreactor. After several weeks of baking and basting, the cells form tissue. Surgeons can then sew the tissue into a patient, whereupon the polymer matrix, or scaffold upon which the cells grow, dissolves away.

A related method involves encasing individual cells in dissolvable capsules, then injecting them into areas where they are needed. The minuscule cages keep the cells secure until they settle down, take hold, and begin to function on their own. Then the cages, too, dissolve away.

Seeking to treat diabetics with such encapsulated cells, William L. Chick, a physician at BioHybrid Technologies in Shrewsbury, Mass., and his colleagues have devised a system for making injectable insulin-producing cells.

Diabetes, which affects an estimated 2.3 million people in the United States, represents the failure of the pancreas to secrete enough insulin to break down sugar circulating in the bloodstream. Diabetics must inject themselves daily with insulin. Chick's treatment strategy involves harvesting insulin-producing cells from pigs, encapsulating the cells in dissolvable "microreactors" — tiny, dissolvable, spherical cages — and then injecting them into the abdomens of people with diabetes, where the microreactors float freely, producing insulin as needed.

"This is a living drug-delivery system," Chick says. "The pancreatic cells have a biochemical mechanism that continuously monitors blood glucose, releasing only enough insulin to keep blood sugar within a normal range. This is a much better system than having to constantly inject yourself with insulin."

The microreactors permit life-sustaining oxygen and nutrients to flow in and wastes and insulin to flow out, keeping cells healthy and nourished. At the same time, they protect cells from antibodies, thwarting an immune reaction. Patients receive an initial injection of cells, to sustain them for 6 months to a year, followed by twice-yearly booster shots. When the cells eventually expire, the microreactors dissolve and the body excretes the waste.

"Using this method, we've successfully treated diabetes in dogs," says Chick, adding that if all goes well, human trials could begin in 2 years. "It's a simple concept. The goal is to make treatment as easy and safe as possible." He also believes that, in principle, the same method could be used to boost the immunity of AIDS patients or to supply the missing blood-clotting factor to control bleeding episodes in hemophiliacs.

Like pancreatic cells, liver cells are finicky, flourishing only under highly favorable conditions. Since a whole liver transplant constitutes the only treatment for end-stage liver disease, which is invariably fatal, David J. Mooney, a chemical engineer at the University of Michigan in Ann Arbor, is attempting to grow implantable liver tissue.

Working with Langer and Vacanti, Mooney has developed a highly porous, spongy material that, when seeded with liver cells, grows into "liverlike tissue."

"The sponges dissolve over time, leaving a natural tissue," says Mooney. "With this technique, we've placed large numbers of liver cells in animals and stimulated blood vessels to grow into the new tissue. The sponge material also contains drugs, which are slowly released, to help the liver cells survive."

The liver of an adult man weighs more than 3 pounds and performs many complicated functions. As the largest visceral organ in the body, it presents "a nontrivial problem" for tissue engineers, says Vacanti.

"We need to get enough tissue transplanted, keep it stimulated, and make sure it has an adequate blood supply and can excrete bile," says Vacanti. "That's a big job."

For people with severely damaged livers, Vacanti would count it a success to replace even one or two of the organ's metabolic functions. "We think we can do that by implanting 10 percent of a whole liver, or about 150 grams of tissue. But that's still a lot of cells."

The research team has succeeded in grafting a healthy portion of liver tissue into several animals. "In Dalmatian dogs, we've shown that implanted cells will replace one liver function for up to 6 weeks," Vacanti says. "That's a good result, though it's still far from ready for clinical testing."

In general engineered structural tissues — cartilage, tendon, ligament, and bone — have come closer to clinical application than visceral tissues such as pancreas and liver. Artificial cartilage has probably made the most headway, with wide applicability in the repair or resurfacing of damaged or arthritic joints and in facial reconstruction for victims of cancer or car accidents. Physicians can mold laboratory-grown cartilage into new ribs, noses, and ears for animals and will soon begin testing in humans.

Efforts to generate new bone have also shown encouraging results. Antonios G. Mikos, a chemical engineer at Rice University in Houston, has cultured bone cells inside three-dimensional polymer scaffolds, then implanted the new tissue into animals. As with other techniques, the polymer provides a structure to help the cells get organized, then dissolves away. Mikos believes this technique will perform well for repairing skeletal defects or replacing missing bone sections.

At Boston's Children's Hospital, researchers have replaced the femur in a rat's leg with engineered bone. The implanted material bears weight, integrates into the animal's skeleton, and holds for an extended period of time. In the case of engineered tendon, the newly grown material forms the right shape but has only one-third of a natural tendon's strength. "We're trying to improve the tendon's mechanical properties," Vacanti says. "But the results here, too, look good.

"We believe there will come a day when we can make an entire composite joint."

For people with severe burns or skin ulcers that will not heal, Naughton's team at Advanced Tissue Sciences is engineering off-the-shelf skin patches for emergency grafts. On thin sheets of dissolvable material seeded with dermal cells, the researchers grow the equivalent of living gauze — coverings for open wounds that stimulate the body's own healing mechanisms. For burn victims, the engineered skin provides a temporary live salve that helps keep in fluids and prevent infection, an alternative to the current practice of using skin from cadavers. For skin ulcers, the patch fills in the gap and promotes wound closure.

In a recent pilot study of diabetics with foot ulcers, 50 percent of the patients treated with engineered skin experienced complete wound closure, compared to 8 percent of the control group, Naughton reports. A follow-up study with 200 patients at 20 centers is expected to begin next year.

"For this population of badly injured people," says Naughton, "the potential benefits from engineered skin are great, while the risks are relatively low. If left untreated, these wounds can lead to infections and even amputation."

To generate enough tissue to treat several hundred thousand people a year, the researchers have filled stacks of bioreactors and tissue banks. The seed cells for the skin grafts come from discarded foreskins following "routine circumcisions," Naughton says. In the laboratory, those hearty infant tissues grow readily into patches.

How much skin have they harvested?

"About 250,000 square feet," says Naughton. "That's roughly the equivalent of six football fields.

"I know," she hastens to add. "We've heard all the jokes."

"People are just beginning to see the potential of this field," Naughton observes. "We're getting close to the point where patients will be able to benefit directly from this technology. Researchers are learning from each other's experiences and developing increasingly complex organs.

"During the next 10 years," she adds, "I expect to see worldwide availability of replacement skin for victims of burns and diabetic ulcers, and possibly of cartilage and blood vessels. Soon after that, we'll probably see whole organs, like livers and kidneys, with active metabolic functions." ❑

****

Questions:

1. What is tissue engineering?
2. How could engineered cells be used as a drug-delivery system?
3. What benefit would be derived from using the body's own cells to grow tissue for replacement?

Answers are at the end of the book.

With all the scientific and political controversy surrounding fetal tissue transplants for Parkinson's disease patients, the search for better treatment continues. Born out of the debate has been the development of methods to improve neuronal transplants using less fetal tissue. How long these fetal transplants can be sustained is still questionable. Excitement is also building over another grafting technique which could eliminate dependence on fetuses altogether. Ideally, the best therapy would be to rescue the patient's own dopaminergic neurons and to prevent their death. Testing has begun on the glial-derived neurotrophic factor (GDNF), a protein which could prevent the death of dopaminergic neurons. Should GDNF work in human patients, it could very well make brain-cell grafts passé.

Researchers Broaden the Attack on Parkinson's Disease

by Marcia Barinaga

Because there is no cure for Parkinson's disease and no therapy that is effective in the long term, researchers are urgently seeking better treatments. Among the treatments now being investigated, the most controversial seeks to replace the brain neurons that die in Parkinson's disease with transplants of fetal tissue. The treatment shows promise, but it is fraught with political and ethical problems because of its dependence on the use of aborted human fetuses.

But four papers published this month suggest that the search for new treatments is broadening out. Two of the papers suggest ways to improve the neuron transplants or make them independent of the availability of fresh fetuses; the other two take an entirely different tack, suggesting that it might be possible to prevent the neurons at risk in Parkinson's from dying by injecting a growth factor into patients' brains. "It is difficult at this point to guess which [of these therapeutic strategies] will eventually turn out to be the most efficient," says neural graft pioneer Anders Björklund of the University of Lund, Sweden. "But one should expect that there will be very interesting developments along these lines in the next few years."

Researchers have been highly motivated to find such new strategies because, despite the large amount of effort put into neuronal transplants during the past 8 years, the therapy has suffered from two major handicaps. Because of the ethical controversy, federal funding of the research was prohibited from 1988 until 1992, when President Clinton lifted the ban. During that period, the research that was done (in other countries or with private funding) gave variable results, with some patients showing tantalizing improvement while others continued to decline.

Many researchers think low survival of the transplanted tissue accounts for at least some of the variability. "Right now what we do is put in as much tissue as possible, and hope and pray that enough cells survive," says University of Chicago transplant researcher John Sladek. The result is that investigators use brain tissue from four to 10 fetuses for each graft.

Better graft survival might reduce the need for so many fetuses, and in a step toward that end, a research team at the University of California (UC), San Diego, led by neuroscientist Fred Gage, recently demonstrated a means of boosting the survival of neurons transplanted into the brains of rats. The group mixed the neurons to be transplanted with nonneuronal cells engineered to produce basic fibroblast growth factor (bFGF), a protein that nurtures many types of neuronal cells in tissue culture. They performed their experiment in rats that had been given a Parkinson's-like condition by injecting 6-hydroxydopamine, a neurotoxin that specifically kills the same set of dopamine-producing neurons that die in Parkinson's disease, into their brains.

When the researchers transplanted the engineered fibroblasts along with fetal dopamine-producing neurons (known as dopaminergic neurons) into the animal's brains, they saw a marked improvement in the survival of the transplanted neurons. Ten times as many transplanted neurons survived in the animals that received the bFGF-producing cells as survived in control animals. What's more, Gage says, the Parkinsonian symptoms were completely reversed in the treatment group, while those of the control rats remained essentially unchanged. Those results, published in the January issue of *Nature Medicine*, represent "the most substantial improvements in cell survival" yet, says Björklund.

It's still too early to tell whether bFGF will end up in clinical use, but researchers are optimistic that it or one of the other growth factors that they plan to test will help sustain the fetal transplants. The growth factors "should have clinical relevance, in the near term, in terms of getting better [graft] survival," says Curt Freed, director of the neurotransplant program for Parkinson's disease at the University of Colorado Medical Center.

Even better, says Björklund, would be finding a way of obtaining dopaminergic neurons that is "largely independent of a continuous supply of fetal material." And that's where the second paper, from a group led by Arnon Rosenthal at Genentech, comes in. Rosenthal and his colleague Mary Hynes were searching for the signals that normally trigger differentiation of dopaminergic neurons during development. They teamed up with Marc Tessier-Lavigne of UC San Francisco, who studies a part of the developing nervous system called the floor plate. "We realized the [dopaminergic] neurons were born near the [floor plate]," says Hynes, and so they wondered whether something in the floor plate might trigger their differentiation.

Their hunch proved correct. In the 13 January issue of *Cell*, they report that floor-plate cells from embryonic rats can induce undifferentiated neuronal precursor cells taken from the rats' midbrain region to become dopaminergic neurons, in both living embryos and the test tube. Even precursor cells that would normally go on to a different fate could be coaxed into becoming dopaminergic neurons instead.

This finding suggests that researchers may be able to grow dopaminergic neurons from undifferentiated precursors in the test tube—which would reduce or even eliminate dependence on fetuses. Working toward that end, Rosenthal says the team is trying to identify the chemical signal in the floor-plate cells that triggers differentiation and is also trying to find a way to grow the precursor cells in continuous culture so they won't have to be collected from fetuses, as they were for the current experiments. The Genentech approach "has the potential to produce almost pure dopaminergic neuroblasts [the nerve cells' precursors] for grafting," says Chicago's Sladek.

Those are encouraging developments in grafting. But many of those gazing into the future of Parkinson's therapy see a time when there will be therapies that rescue a patient's own dopaminergic neurons, making neuron grafts unnecessary. One candidate as a substance for preventing the death of dopaminergic neurons is glial-derived neurotrophic factor (GDNF), a protein isolated in 1993 by Frank Collins and his colleagues at Synergen Inc., a biotechnology company in Boulder, Colorado. In work reported in two papers in this week's issue of *Nature*, Lars Olson of the Karolinska Institute in Stocholm, Barry Hoffer of the University of Colorado and their colleagues, and Franz Hefti and Klaus Beck of Genentech and the co-workers show that direct injections of GDNF into the brains of rats and mice can save dopaminergic neurons that have been damaged by either the neurotoxin MPTP or surgical injury.

In the Genentech group's experiments on surgically injured rats, the control animals lost 50% of their dopaminergic neurons, while animals that were treated with GDNF after the injury retained nearly normal numbers of healthy cells. Hoffer's group had similar results with mice treated with MPTP, and in addition found that the GDNF injections also alleviated the Parkinson's-like symptoms induced by MPTP. (The neuron-severing injury used by Hefti's group doesn't produce symptoms in the rats, so they couldn't test for recovery.)

Although preliminary, these experiments have raised hopes that GDNF will have similar neuron-saving effects in humans. But Freed cautions that it's not yet clear whether results can be extrapolated to humans, because the existing animal models are only approximations of Parkinson's. "We simply don't know," he says, what effects growth factors such as GDNF might have in human patients.

We might know soon, though. Amgen Pharmaceuticals bought Synergen last month and plans to move toward clinical trials, although Collins, now at Amgen, says there is no schedule yet for the trials. The first trials will probably involve pumping GDNF directly into the patients' brains, because it is a protein that does not cross the blood-brain barrier. If GDNF proves successful, Amgen and other companies will no doubt search for a small molecule that mimics the protein's effects but can cross the blood-brain barrier and therefore can be administered systemically.

Will this therapeutic approach eventually make brain-cell grafts obsolete? "It is too early to call it," says Hefti, because the approaches to both transplantation and rescuing the patients' own neurons are in such early experimental stages. "Both approaches are very valuable at this point," he says, "and they should be pursued." And it's possible that one of them will prove to be a much better substitute for the existing therapies and their merely temporary respite from the devastation of the disease. ❑

****

Questions:

1. What accounts for the variability of fetal transplants?
2. What would be the advantage of growing dopaminergic neurons from undifferentiated precursors in continuous culture?
3. Why does GDNF have to be injected into the brain?

Answers are at the end of the book.

There are an estimated 1.5 million Americans who are afflicted with Parkinson's disease. A slow path to destruction begins with a tiny section of the brain responsible for supplying the neurotransmitter dopamine to the center of the brain which controls movement. Without dopamine, the body's movements slow down to a snail's pace requiring the patient to become confined to a wheelchair. For some people with Parkinson's, their only treatment is a highly experimental and controversial technique — a fetal tissue transplant. The procedure is so experimental that its effectiveness has never been fully documented. The source of the tissue, brain cells from aborted fetuses, caused such political fury that the government banned testing the technique in humans during Reagan and Bush presidencies. To overcome these obstacles, researchers are developing alternative sources of tissue, like the patient's own skin and muscle cells, genetically engineered to produce the necessary compounds the brain needs to make dopamine. Until progress is made, fetal tissue transplants remain the only hope to patients with Parkinson's as the scientific and political battles rage on.

Fetal Attraction

In theory, brain cells that have been killed by Parkinson's disease can be replaced with cells from the brains of aborted fetuses. Now that the necessary politics and the technology are in place, neurosurgeons are about to find out if that theory is correct.

by Jeff Goldberg

Over the past 11 years, as her Parkinson's disease has progressed, 68-year-old Thelma Davis has come to feel trapped in a body that will not move. The symptoms were unalarming at first. Davis told herself that the slight limp in her left leg was nothing serious, that the weakness in her left arm was just her imagination. But although the early signs of Parkinson's disease are so subtle that they are often ignored by patients and misdiagnosed by doctors, the disease takes a relentless course. Uncontrollable tremors began to appear in Davis's hands and legs. Because the disease affects movements of the jaw and mouth, speech became difficult for her. Little by little, her gait slowed to a flat-footed shuffle, and her face froze into the unblinking, unsmiling mask characteristic of Parkinson's sufferers.

The cause of the disease, which afflicts an estimated 1.5 million Americans, remains unknown, and there is no cure. Scientists do know that Parkinson's casts its imprisoning spell by slowly destroying a tiny section of the brain, the size and shape of a quarter, called the substantia nigra. The substantia nigra supplies the neurotransmitter dopamine to a larger area in the center of the brain, the striatum, which controls movement. As dopamine supplies from the substantia nigra to the striatum dry up, movements slow, become erratic, and finally grind to a halt.

Parkinson's patients, like the rest of us, have plenty of dopamine elsewhere in their bodies; the conundrum is how to get it into their brains. Dopamine can't pass through the blood-brain barrier, a membrane that guards the interior of the capillaries in the brain. Fortunately, the drug levodopa, commonly known as L-dopa, can pass through the barrier, and once it reaches the substantia nigra, cells there convert it into dopamine.

Thanks to L-dopa, Davis was able to lead a relatively normal life for several years, continuing to work as the chief financial officer of a Long Island, New York, mortgage bank. But L-dopa inevitably fails as Parkinson's destroys substantia nigra cells, eventually leaving too few to convert the drug to the neurotransmitter. When her symptoms worsened, Davis reluctantly retired from her job. Now, despite a three-times-a-day regimen of short-acting and timed-release forms of L-dopa, she finds simple tasks like combing her hair and dressing difficult obstacles. She suffers from episodes of "freezing" when the drugs wear off, alternating with spurts of convulsive herky-jerky movements when they shock her system into overdrive — the classic "on-off" symptoms of advanced Parkinson's disease. Parkinson's itself is not fatal, but many patients die from injuries suffered in falls. Others end up wheelchair bound, unable to move or speak, or succumb to pneumonia.

With her condition deteriorating, Davis has come to the Neural Transplantation Center for Parkinson's Disease at the University of Colorado in Denver for what could be her last hope of recovery — a fetal tissue transplant. In the operation, transplant team leader Curt Freed and neurosurgeon Robert Breeze will implant brain cells culled from aborted fetuses through a thin needle into Davis's brain. To make sure the dopamine reaches the cells that need it, the fetal tissue is grafted into the striatum, where neurons are alive but deprived of dopamine, rather than the substantia nigra, where they're dying. Davis and her doctors hope that as the grafted cells grow and integrate into her brain, they will pump out enough dopamine to replenish depleted supplies and give her back some of her lost mobility.

The tissue for the transplant has been collected, with the consent of the mothers of the fetuses, from private clinics where abortions are performed. It consists of brain cells from the mesencephalon, an area that develops into the substantia nigra and other midbrain structures, dissected from half a dozen six-to-eight-week-old fetuses. The cells must be collected within a narrow window that opens between six and eight weeks into the gestation of each fetus, just before they have fully differentiated into dopamine-producing neurons. "Brain cells at this age can grow just like seeds," says Freed. "They establish root systems in the form of neural connections," regenerating damaged brain circuits. (If the cells are any older, they break up and die during the transplant process.) Because these fetal cells have not yet developed the antigens that trigger an immune response, they also appear to grow without rejection. Over the past 20 days, the cells have been cultured, screened for bacteria and infectious diseases, and tested for levels of dopamine production. Twice during the last three months, Davis's operation has had to be delayed because the tissue was less than perfect.

Davis begins her day by having a metal band bolted to her skull by four pins embedded in the outer layer of bone. A device that looks like a delicate geodesic dome is attached to the band, preparing Davis for her magnetic resonance imaging (MRI) scan. As a gurney inches her forward and back through a powerful doughnut-shaped electromagnet, a scanner detects radio signals emitted by hydrogen atoms in her brain. These signals are reassembled into three-dimensional images by the machine's computer and projected onto a bank of monitors, which Breeze studies intently. From the images, Breeze calculates the angles and routes of the needles that will insert the fetal tissue implants into target sites while avoiding injury to arteries and vital brain structures. The domelike device provides a grid of reference points for plotting these routes.

After Davis is prepped for surgery and wheeled into the operating room, the dome is replaced with a stereotaxic frame, an awkward-looking device that resembles a large compass or sundial. The frame is a precision measuring tool. Its outer rim contains an array of small holes that can be adjusted to within a fraction of a millimeter to guide the needles delivering the fetal tissue.

Breeze drills four holes a shade smaller than the diameter of a pencil into Davis's forehead and through her skull. As he carefully inserts the needle, Freed prepares the tissue, which has been transported to the operating room in a blue-and-white cooler. The tiny specks of tissue are suctioned into a syringe designed to extrude the tissue in fine, noodlelike strands. These are loaded into a hollow stainless steel tube called a cannula.

The operation is nearly bloodless and, since brain tissue does not register sensation, almost painless as well. Davis is sedated with a local anesthetic but remains fully awake. During stereotaxic procedures (which are most often used to obtain biopsy specimens of suspected brain tumors), it's better to keep patients awake and talking, Breeze believes, to help guard against even the remote chance that the needles and catheters inserted into the brain could cause bleeding and precipitate a stroke. While general anesthesia would routinely be used for open brain surgery, when Breeze can see what he's doing, during stereotaxic surgery he works blind, directing instruments into the brain based on computer calculations alone. If the patient were asleep and her brain began to bleed, by the time the doctors noticed, it could be to late. So the anesthesiologist keeps up a steady conversation with Davis throughout the operation, carefully listening for any confusion or slurring of speech.

The hollow needle is equipped with an inner stylet, to make it a solid probe that will not cut a core from her brain as Breeze taps the device gently forward. When the needle is in place, Breeze removes the stylet and replaces it with one of the cannulas that Freed has filled with fetal cells, and the infusion begins.

Two hours later, with two small bandages covering the incisions in her forehead, Davis is wheeled out of the operating room into intensive care. She'll go home after four days, but it will be months more before she knows whether the transplant has worked. Fetal tissue transplantation for Parkinson's disease remains highly experimental, and Freed cannot promise a positive outcome.

Thelma Davis is only the twenty-second patient to undergo the procedure in Denver. Fetal tissue transplants for Parkinson's disease are also offered on a limited basis at Yale, the University of South Florida in Tampa, and the Good Samaritan Hospital in Los Angeles, as well as in England, France, and Sweden, where some of the first experiments with the procedure were performed in the mid-1980s.

Roughly 200 such operations have taken place worldwide since the technique was introduced amid a storm of political controversy over the source of the transplanted tissue — electively aborted fetuses. In this country, despite the judgment of a National Institutes of Health advisory committee that fetal tissue transplantation was ethical and promising, government funding to test the technique in humans was banned during the Reagan and Bush presidencies. As a result, clinical studies were limited to a handful of patients. That ban was lifted by executive order during the first days of the Clinton administration, but now a new debate has surfaced, this time among scientists, over how well the transplants work.

"The moratorium distorted the scientific discussion," says D. Eugene Redmond, the leader of a transplant team at Yale. "To muster the political power to overturn it, the actual scientific accomplishments were somewhat exaggerated."

"There was a presumption that it would work if the ban wasn't there," adds William Freed (no relation to Curt Freed), an NIH researcher. "People thought, 'Well, it's banned; it must be something really great.' "

Some patients have shown marked improvement on standard movement tests, such as touching a thumb and forefinger together or tapping their feet, and have resumed many daily activities that most people take for granted: tying their shoes or their ties, vacuuming or driving. They are able to reduce their medication by an average of 50 percent. One of Curt Freed's patients, a California telephone lineman who had nearly lost his ability to speak and was embarrassed to eat with friends because he could no longer feed himself properly, celebrated the one-year anniversary of his transplant with a Thanksgiving dinner for 12. Another now enjoys cross-country skiing. "About a third of the patients have had their lives revolutionized," says Freed. "The problem is, the effects are variable."

One in three patients shows only moderate gains, and another third experience no long-lasting benefit at all from the operation. A few even get worse. There are other risks as well. The chance of something going catastrophically wrong during a transplant procedure is small — less than 1 percent that a needle will inadvertently strike an artery or a vital brain area. Nevertheless, in January 1994, Freed's seventeenth patient, a 55-year-old man with an eight-year history of Parkinson's, suffered a stroke in the operating room and died one month later — the first procedure-related fatality.

"We knew this was an odds game," says Freed. "Passing needles into the brain carries risk, and the risk of stroke is about 1 in 500 needle passes. At the time, we'd been doing 14 to 16 needle passes on each patient. With each operation, there was about a 3 percent chance of stroke."

Davis and other patients, many of whom have paid for the $40,000 operation privately, are willing to take that chance. "No other form of treatment holds as much hope as this," says Davis. Neurologists who routinely care for Parkinson's patients remain cautious about recommending the procedure without more proof, however. "Parkinson's disease is slowness of movement, not paralysis," points out Stanley Fahn, a Parkinson's specialist at Columbia Presbyterian Medical Center in New York. "Sometimes with enough excitement or stimulation, sudden movement can return. But this is only transient." Patients in advanced stages, who freeze when they try to cross doorways or are unable to walk across a room without holding onto the walls and furniture, can often negotiate a flight of stairs with ease, ride a bicycle, or catch a ball.

Fahn worries that some improvements observed in transplant patients may be the result of the excitement of undergoing the operation, the mystique of the transplant procedure, and the expectation of getting well. Parkinson's patients may be particularly prone to such placebo effects. Studies of new drugs have shown that as many as 30 percent of Parkinson's patients improve with placebo medications, albeit only briefly. Similarly, neurosurgical answers for Parkinson's disease are also suspect.

In one such surgical technique, called pallidotomy, surgeons destroy a minuscule area in the movement center of the brain — the internal globus pallidus — which is located at the base of the brain, just above the spinal column. The procedure was recently reported on *Prime Time Live* to reduce tremor dramatically in Parkinson's patients. But the *New York Times* followed up with a story detailing how the positive effects may not last, while the operation often leaves patients worse off than before. Skeptics also point out that tremor is only one among many symptoms of Parkinson's. Before L-dopa became a standard treatment, lesion therapies, in which surgeons destroyed parts of various brain structures (usually the thalamus), were also observed to relieve Parkinson's tremors, but these operations didn't relieve slowness of movement in any lasting way.

Autologous transplants, using dopamine-producing cells from a patient's own adrenal glands, also proved to be a disappointment. In the late 1980s hundreds of adrenal transplants were performed throughout the world (including about 100 in the United States) after Mexican neurosurgeon Ignacio Madrazo reported startling successes with the procedure. About 40 percent of the patients did experience some initial positive effects, but the benefits of the operation generally vanished before a year had passed. At least part of the problem, according to Freed, was that these cells produce mostly epinephrine and norepinephrine, with only a little dopamine. "They're just not the right kind of cell," he says. "Also, they don't survive well in the brain, because they don't belong there. The tissue around them doesn't provide a supportive environment."

The results suggest the patients were experiencing a placebo effect, but the side effects of the open brain surgery were real enough: respiratory problems, pneumonia, urinary tract infections, sleeplessness, confusion, and hallucinations. Patients had also suffered strokes and heart attacks while undergoing the surgery; about one in ten patients died.

To put the effectiveness of the fetal tissue operation to the test, Freed and Fahn have joined forces on an unprecedented and controversial study. Forty patients, screened by Fahn at his clinic in New York, will undergo transplants in Denver. But to assure that whatever improvements the patients enjoy can legitimately by attributed to the procedure, the study will follow a double-blind, randomized design, similar to a drug trial, in which half the patients will receive a sham operation. Researchers will give the control patients an MRI scan, prep them for surgery, fit them with stereotaxic frames, and drill holes in their skulls; then Breeze will fake the rest of the procedure. Neither the patients nor Fahn, who will be evaluating their progress, will know who has actually received an implant.

"The pacing and atmosphere will be nearly identical to the true tissue implant," says Freed. "The strategy is to do things exactly the same way, maybe even have some tissue set up in a dish so there's time involved in picking up the tissue. We'll drill holes in the skull, the needles will be inserted into the stereotaxic apparatus, all the calculations will be done, the timing will be exactly the same, but the needles will not drop the last 5 centimeters into the brain."

After the operations, Fahn will evaluate the patients by methods ranging from rating their performance on tasks like getting out of a chair to computerized analysis of their videotaped movements. Researchers will also perform positron-emission tomography (PET) scans to evaluate how well the tissue has survived and grown in the patients.

Members of the control group will be eligible to receive real transplants later — providing the procedure passes the test. "We've promised them the treatment. But if it's a bust, they're better off having the sham surgery rather than the real operation." says Fahn.

The four-year, $4.8 million trial is being funded by the National Institutes of Health. This is the first grant awarded for a study of fetal tissue transplantation since the research moratorium ended. A second $4.8 million NIH grant has recently been approved for a similar controlled study of 36 patients who will undergo transplants at the University of South Florida in Tampa.

Freed hopes the studies will provide an unbiased estimate of the value of the procedure. The sham operation presents no additional risk, he says. As an added precaution, the NIH has assigned a Data Safety Monitoring Committee to oversee the studies. The committee has the option of stopping the studies if there are signs of any unexpected complications.

"What you don't want to do, especially with something as dramatic and publicized as fetal tissue transplantation, is put yourself in a position where you're not sure that what you're seeing is real," says C. Warren Olanow, a neurologist at Mount Sinai Hospital in New York. Olanow heads the consortium of investigators that will conduct the second NIH-funded trial. "Is it better to expose a small number of patients to placebo than to forgo a control group and potentially expose hundreds of thousands to a procedure that may not work?"

However, other researchers consider the double-blind studies dangerous and therefore unethical. "There's a one in a hundred chance that performing craniotomies on the surgical controls could result in the formation of blood clots. If one of those patients dies, it could set the field back several years," argues neuroscientist John Sladek of the Chicago Medical School.

Controlled trials of fetal tissue transplantation will remain premature, at best, critics believe, until gaping methodological differences between the transplant teams are resolved. The teams disagree on such crucial details as in which of the two sections of the striatum — the putamen or the caudate nucleus — to implant the tissue. The putamen appears to be responsible for a wider range of Parkinson's disease deficits, such as freezing and the inability to walk, but the caudate may govern a number of subtle functions, including eye movement. Researchers also disagree about how much tissue is needed (one to nine fetuses), how to prepare it for transplant (in suspensions, cultures, or cryogenically frozen and thawed specimens), and whether to place large quantities in a few locations or skewer the brain with 15 to 20 micrografts. Even the key question of whether patients should be treated with the immunosuppressant drugs used in organ transplant operations remains undecided.

Improved tissue processing and implantation techniques could also resolve another major concern of researchers, the poor survival of fetal tissue. Recently, when one of Olanow's patients died (of unrelated causes) a year and a half after his surgery, an examination of his brain revealed that hundreds of thousands of fetal cells had survived and formed connections with surrounding brain tissue. But a handful of other autopsy reports and numerous animal studies, showing that up to 95 percent of the transplanted cells die, reinforce the need for further progress. "Although even a few surviving transplanted cells may be enough to produce clinical effects, poor survival may account for variability in the results we've seen so far," says Freed. "People doing kidney transplants were able to get better results simply by improving their surgical techniques and their handling of the organ."

Another problem is the needle-in-a-haystack task of gleaning usable dopamine-producing cells from 2-centimeter-long fetuses often smashed beyond recognition during abortions. The difficulty, Freed once told a Senate subcommittee, is so great that it should be an adequate safeguard against any potential abuses of fetal tissue transplants. Problems with tissue availability may also continue to limit the number of fetal tissue transplants that can be done in Denver and other centers.

To overcome these obstacles, researchers elsewhere are exploring a number of new technologies, including new versions of autologous transplants. This time the cells used would be the patient's own skin and muscle cells, genetically engineered to produce tyrosine hydroxylase, a chemical the brain uses to make dopamine, as an alternative source of tissue. "From a skin biopsy the size of a quarter we can produce as much tissue in two weeks as you could harvest from a hundred fetuses," says Krys Bankiewicz, who is working with the California biotechnology company Somatix Therapy to perfect methods of mass-producing cells for transplant.

In April, doctors at the Lahey Hitchcock Clinic in Burlington, Massachusetts, began trials of cross-species transplants, inserting tissue from five fetal pigs into the brain of a Parkinson's patient. Porcine fetal brain cells are similar to human fetal cells but are more readily available. Experiments are also under way to test "encapsulated" dopamine-producing cells sealed in semipermeable plastic capsules. Because this delivery system is designed to allow dopamine out of the cells while preventing immune destruction, scientists hope unmatched adult tissue or even animal tissue could be transplanted into patients without immune suppression.

But dopamine may not be the whole story. Some researchers believe the fetal transplants may also produce growth factors — chemicals that stimulate nerve cells to sprout. In a recent article in the *Journal of Neurosurgery,* Bankiewicz describes experiments he conducted at the NIH in which he transplanted a variety of fetal cells — none of which produce dopamine, but all rich in growth factors — into rats and monkeys. The result was nearly as good as fetal transplants of mesencephalic tissue, producing "a measurable improvement" in the animals lasting 7 to 12 months.

According to Bankiewicz, the healing powers of fetal tissue grafts stem from a "dual effect" of dopamine and growth factors. Future studies may include trying to boost the effectiveness of fetal tissue grafts by infusing additional growth factors or injecting growth factors alone.

For the time being, however, patients seeking a transplant to relieve their Parkinson's disease have only one option — fetal tissue. "The field is still in its early stages. But I'm very optimistic that as our techniques improve we will have a chance of curing Parkinson's," says Freed. "The patients are desperate, and we have no other means of helping them. If they could wait five years, we could probably do it better. But some can't."

****

Despite the risks and unknowns, there's no lack of volunteers for the trials. Enrollment has been completed in the 40-patient Denver study, and though the 36 patients for the Tampa trial have not yet been selected, Olanow has had no trouble finding willing subjects for the study's preliminary stages. The Denver patients range in age from their early fifties to 75; Olanow expects the Tampa patients to start as young as 35. In both studies half the patients will be under 60 to see whether the patient's age alters the effectiveness of the procedure.

Olanow warns his patients that they must have realistic expectations. "We need people who are absolutely committed to seeing it through. If a patient is unrealistic and he doesn't have a great result, you won't see him again. We need to be able to follow these patients, especially the bad results, because you've got to know what went wrong as well as what went right."

For Thelma Davis, as she recovers at her home on Long Island after the long flight back from Denver, waiting to see if the transplant worked is going to be tough. "You try to rationalize the situation, the facts, that it's going to take months and may not be a cure. But it's not logical," she says. "I just hope I have the patience." ❑

Questions:

1. What part of the brain is destroyed by Parkinson's disease?

2. What is L-dopa?

3. What are autologous transplants?

Answers are at the end of the book.

Part Three:

The Brain, The Peripheral Nervous System and Properties of Neurons That Serve Them Both

How we move our muscles in a smooth coordinated fashion is directed by the primary motor cortex of the brain. It may sound simple and it appeared so to Wilder Penfield who in the 1930s drew the first motor cortex map mirroring the body plan. But the organization of the brain is not as simple as one cortical area controlling specific body parts in an orderly fashion. This idea is changing and it is now described by neuroscientists as a complex mosaic of neurons controlling many different body parts. With more sophisticated imaging techniques, current research is finding muscle activity or movement to be coordinated by certain neurons to produce a common component of many movements while a second group of neurons would provide fine motor adjustments when needed. This arrangement allows cortical neurons to make connections with many different muscles to accomplish a task as seemingly simple as tying your shoe.

Remapping the Motor Cortex

The primary motor cortex of the brain does not contain an orderly map of the body but is instead a complex mosaic of neurons controlling different body parts

by Marcia Barinaga

When you're at the gym, sweating your way from one weight station to the next, you may be focused on bulking your biceps or strengthening your pectoral muscles. But in spite of your concentration, even the fanciest weight machine can't work just a single muscle at a time. Multiple muscles must cooperate to choreograph even the simplest movements made by anyone, from an Olympic champion to a couch potato pushing buttons on the remote control.

Such movements do not, of course, begin with the muscles but must be orchestrated by the brain and spinal cord. Recent work from numerous labs including that described by neurobiologists Jerome Sanes and John Donoghue of Brown University and their co-workers in Steven Warach's group at Harvard Medical School is now clarifying just how a brain region known as the primary motor cortex helps coordinate muscles to produce a movement.

The findings, which began over a decade ago but have intensified in the past few years, overthrow an earlier view that the primary motor cortex contains an orderly map mirroring the body plan, with a one-to-one correspondence between cortical areas and the body parts they control. Instead, there is now evidence of a complex and disorderly mosaic in which multiple cortical areas control a given body part. "An individual muscle or movement, or part of the body ... is represented diffusely, and it is intermingled with the representation of lots of other parts," says University of Rochester neuroscientist Marc Schieber.

Despite its disorderly appearance, this evolving picture of the primary motor cortex provides some intriguing insights into how this brain area may function. The diffuse pattern in which the cortical neurons are wired to muscles may make them ideally suited to select, from a nearly infinite set of possible muscle combinations, the unique combination required to produce any movement. "It could be that inherent in this very complicated, messy-looking patchy mosaic map are solutions to many of these motor control problems," says neuroscientist John Kalaska of the University of Montreal.

The idea of a map in the primary motor cortex dates to the work of Canadian neurosurgeon Wilder Penfield in the 1930s. Penfield electrically stimulated the brains of epilepsy patients during surgery and found that the cortical neurons whose activity caused movement of the various body parts are arrayed in an orderly way that roughly reflects the body plan. The foot and leg lay at one end of the map, followed by the torso, arm, and then disproportionately large areas devoted to the most agile body parts, the hand and face. Penfield depicted his findings in the form of a drawing of a distorted little man, or homunculus, draped across the surface of the cortex.

Some neuroscientists have interpreted Penfield's homunculus in the most literal way — assuming, for example, that there are side-by-side patches of cortical neurons that govern each of the five fingers of the hand — but most agree that Penfield didn't intend that interpretation and instead only meant to show that there are broad areas of primary motor cortex devoted to governing the leg, torso, arm, and head. At that level of resolution, Penfield has turned out to be correct. But it is the next level of resolution, the issue of how neurons are organized within those areas, that more recent research has begun to unravel.

3D movements a problem

One of the main conundrums the research has had to address is the fact that something as complex as muscle activity or movement is not readily mapped onto a two-dimensional sheet of neurons such as the primary motor cortex. Consider the elbow joint, which is controlled by many muscles working in different combinations to produce different movements. What is more, the elbow doesn't move in isolation; movements of the elbow put force on the shoulder and wrist as well, so other muscles must be activated to stabilize those joints. "You would be expecting too much if you thought you could represent three-dimensional movements and muscles on a two-dimensional cortex and have a neat point-to-point map," says Rochester's Schieber.

Neuroscientists considering this problem decades ago hypothesized a simple solution: They suggested there might be a one-to-one correspondence of cortical cells to individual muscles. But work done in the 1960s and 1970s laid that notion to rest by showing that individual muscles get input from multiple spots on the cortex and individual cortical neurons have branches connecting them to motor neurons that control more than one muscle. "The relationship was not one-to-one, but one-to-multiple," says Yoshikazu Shinoda of the Tokyo Medical and Dental University, one of the researchers involved in this early work.

With individual cortical neurons linked to multiple muscles and each muscle in turn linked to multiple spots on the cortex, researchers began to wonder if there was a different kind of rhyme or reason to the wiring. Perhaps the muscles driven by an individual neuron might work in concert to produce a given movement.

Studies that began in the late 1970s and are still going on today have shown that such a correspondence does exist. This work got under way when Paul Cheney and Eberhard Fetz, at the University of Washington, Seattle, developed a means of correlating the activity of single cortical neurons with the activities of muscles used by monkeys in performing trained movements. When they used the method to study simple wrist movements, they found that certain individual neurons controlled muscles that worked together in a synergistic way, producing wrist extension, for example. Those same neurons also inhibited muscles producing the opposite movement. "This is a good example of functional synergy that is well-suited for producing movement in a particular direction at one joint," says Cheney, who still studies muscle synergies at the University of Kansas Medical Center in Kansas City.

The synergy discovered by Fetz and Cheney was not limited to just one joint, as other researchers showed. Roger Lemon, at Cambridge University, UK, applied Fetz and Cheney's technique to the fingers, whose amazing dexterity led many researchers to argue that they, more than any other body part, must be controlled by a one-to-one neuron-to-muscle correspondence. But Lemon reported in the mid-1980s that he hadn't found such a fit. "Most of [the neurons] address a number of different muscles," says Lemon. "They also span muscles activating at the wrist with muscles that activate the fingers."

Then, just last year, Brian McKiernan, Jennifer Karrer, and their co-workers in Cheney's lab expanded the notion of synergy even further. McKiernan reported at last November's meeting of the Society for Neuroscience that when they studied monkeys doing a task that involves reaching forward with one arm — a movement that employs wrist, elbow, and shoulder — they found individual neurons that appear to control muscles at all three joints. In their study, multi joint connections weren't a fluke; nearly half the neurons the team sampled were linked to muscles at more than one joint.

The bottom line is that many cortical neurons control synergistic sets of muscles, and most muscles are controlled by multiple sets of neurons. But those findings alone don't rule out the possibility of an orderly, albeit somewhat fuzzy, map. For example, within Penfield's arm zone, Cheney says there seem to be "separable regions for distal muscles [that control the wrist and hand] and proximal muscles, proximal being elbow and shoulder." Such zones were in fact identified in the arm zone of the primary motor cortex by John Murphy and his colleagues at the University of Toronto in the 1970s. They found a cluster of neurons controlling the hand area, with neurons controlling the more proximal areas of the arm arranged in concentric horseshoe shapes around that core.

Mixing it up in the cortex

That horseshoe-shaped map may not resemble the arm of the homunculus, but it nevertheless constitutes a map, with contiguous areas representing the parts of the arm. But the idea of individual contiguous areas was called into question in 1978, when Peter Strick and James Preston, at the Veterans Administration Medical Center and the State University of New York Health Science Center at Syracuse, found that the primary motor cortex contained a second cluster of hand-specific neurons.

Strick and Preston's finding foreshadowed the further fragmenting of the map that was to come. Several studies published within the past 5 years suggest that the arm area is in fact riddled with zones that specialize in hand control, intermingled and overlapping with other zones that control the upper arm. In the early 1990s, for example, Sanes and Donoghue at Brown, with Steven Leibovic at the Massachusetts Institute of Technology, measured the responses of 14 hand and arm muscles in monkeys while electrically stimulating neurons all over the arm area of the motor cortex. They found a mosaic of cortical patches that control hand muscles, separated by other patches that activate muscles of the upper arm. There was also plenty of overlap between areas controlling the arm and those controlling the hand; it was in one of those overlap zones that Cheney's group found the neurons that control muscles of the wrist, elbow, and shoulder.

In 1993, Schieber, then at Washington University in St. Louis, and Lyndon Hibbard, also of Washington, took a different approach to the same problem. Reasoning that the artificial electrical stimulation applied by Sanes and Donoghue might not reflect normal cortical activity, they instead recorded the activity of primary motor cortex cells while monkeys were making trained movements of different fingers. Like Sanes and Donoghue, they found that widely spaced sets of neurons are involved in each finger movement. Although the set for each finger was different, there was extensive overlap among the neurons included in any one set. That, says Schieber, suggests that "the control of whether the thumb moves or the little finger moves is not handled by a spatial map," but instead depends on the activity pattern of a whole population of neurons.

Researchers study monkey motor cortex because they can't stick electrodes into peoples' heads, and monkey brains are the closest approximation to our own. But noninvasive imaging studies are confirming that the human motor cortex indeed resembles that of monkeys. Studies using positron emission tomography (PET) in several labs have suggested that the areas controlling arm and hand movements are interspersed with one another in the primary motor cortex.

And now two studies, one by Sanes and Warach and their co-workers in this issue, and another by Stephen Rao of the Medical College of Wisconsin in Milwaukee and his collaborators, published in the May issue of *Neurology*, have applied functional magnetic resonance imaging, which has a higher resolution than PET, to the problem. These results show even more convincingly that individual finger movements are apparently controlled by networks of neurons that occupy overlapping areas within the primary motor cortex. "The pattern that you see is unique for each hand movement," says Sanes, but the different patterns "overlap by about half."

The picture emerging from all this work is one in which neurons scattered throughout a given large body zone of the primary motor cortex (such as the one representing the arm) can be called on to activate the multiple muscles necessary for a movement. And that, says

Lemon, now at Queens Square Hospital, London, suggests that rather than a spatially organized map, there is an entirely different kind of map, "a map of combinations of muscles, arranged in useful ways, almost like books on a shelf that you can pull out when you need them." In such a scheme certain neurons might control a coordinated set of muscles to produce a common component of many movements, such as reaching forward with the arm, while others would be added to provide the fine-tuning necessary to make that movement unique.

Getting it together

But if those neurons are scattered around the arm area, what provides the look-up system, as it were, to pull the right books off the shelf? Sanes points to a 1991 study by George Huntley and Edward Jones, at the University of California, Irvine, that he thinks may provide a clue. Jones and Huntley showed that an extensive network of connections links neurons throughout the arm area of the primary motor cortex, but doesn't cross over into the neighboring areas that control the face and torso. That, says Sanes, is just the kind of pattern that would be required to connect the scattered neurons involved in arm movement.

Such connections would reduce the need for an orderly map, adds Schieber, because neurons could be called into action in a way analogous to how computers store files on a disk. "The computer can write different bits and pieces of one file at different spatial locations on the disk," he says, "then it goes from place to place on the disk, picking up pieces of information it wants."

Such a dispersed system would be well suited to providing plasticity — the constant reforming of neural liaisons associated with learning, which is known to go on in the motor cortex as a person or monkey learns new motor skills. "The horizontal connections ... would allow new representations to emerge and to come back to their original forms in a very dynamic way," Sanes says.

The new findings also provide clues to a special function that primary motor cortex may play in movement control. The cortex is not absolutely necessary for movement; other brain areas also make connections with the motor neurons of the spinal cord. Cats can walk even when their cerebral cortex has been removed, and monkeys without a primary motor cortex can be trained to make many types of movements. Such findings raise the question of what essential tasks the primary motor cortex performs.

SUNY's Strick proposes that the multi-muscle connections of cortical neurons make the primary motor cortex particularly suited to guide complex actions that require the coordination of several muscles. In support of this hypothesis, Strick and co-worker Donna Hoffman have shown that monkeys with lesions in their primary motor cortex, while severely handicapped at first, can be rehabilitated to do many simple movements that depend largely on single muscles, such as flexing and extending their wrists. As Strick and Hoffman found earlier this year, however, the animals never regain their ability to make smooth diagonal movements that use two muscles simultaneously, such as wrist flexion with deviation to the side. To make that diagonal movement, Strick says, the monkey is reduced to using first one muscle, then the other, producing "zigzag movements, as if tacking like a sailboat." That, he adds, suggests that "moving on the diagonal is something that would require the unique branching patterns of the cortical neurons."

So out of the remapping of the primary motor cortex comes not only a new understanding of how this brain area is arranged, but also new insights into its crucial role in complex movements. And that means that whether you are pumping iron or doing something as routine as tying your shoes, you can thank your primary motor cortex that you can glide smoothly through the movements without any zigs or zags. ❑

Additional Reading

M. H. Schieber and L. S. Hibbard, "How somatotopic is the motor cortex hand area?" *Science* **261**, 489 (1993).

J. P. Donoghue, S. Leibovic, J. N. Sanes, "Organization of the forelimb area in squirrel monkey motor cortex: Representation of digit, wrist and elbow muscles," *Experimental Brain Research* **89**, 1 (1992).

D. S. Hoffman and P. L. Strick, "Effects of a primary motor cortex lesion on step-tracking movements of the wrist," *Journal of Neurophysiology* **73**, 891 (1995).

G. D. Schott, "Penfield's homunculus: A note on cerebral cartography," *Journal of Neurology, Neurosurgery and Psychiatry* **56**, 329 (1993).

****

Questions:

1. What region of the brain helps coordinate muscles to produce movement?

2. Why are most movements not controlled by a one-to-one neuron-to-muscle correspondence?

3. What advantage is there in having extensive neural networks scattered throughout an area of the cortex?

Answers are at the end of the book.

Developmental neurobiologists have known for years that the nervous system doesn't work unless it is wired properly. One important question being asked is what makes it possible for neurons to find their way through a maze of tissues to arrive at their targets several feet away. Mystifying as it sounds, neuronal target cells can secrete special proteins called chemoattractants which when released into surrounding tissue can guide the correct axons to their destinations. Those molecules that contribute to the attraction of axons belong to a family of proteins named netrins. Likewise, complementing the secreted attractive factors that exist are proteins with chemorepellent activity called semaphorins. Since their discovery, scientists have been particularly interested in how these diffusion proteins work to guide the axons of the spinal cord which relay information about pain and temperature to the brain. Although it is too early to say definitively which neurons respond to which guidance molecules, with continued progress, scientists hope that someday they will be better able to treat the more serious nerve injuries of the spinal cord.

Helping Neurons Find Their Way

A flurry of recent papers suggests that the growing axons of neurons are guided to their targets by diffusible chemorepellents as well as by attractants

by Jean Marx

If you've ever been reduced to despair by a last-minute Christmas Eve effort to assemble a bicycle as a gift for one of your children, take heart. Your assignment could be worse. Much worse. Try assembling the nervous system of a developing embryo. While you would have more than one frantic evening to accomplish the job, in higher organisms the task requires making billions of precise connections between nerve cells, as well as between nerve cells and muscles and other body tissues. Often the axons, the nerve cell projections that make the connections, must travel distances of up to several feet — cosmic distances for cells — through a maze of many tissues to find their targets. And, as with bicycle assembly, the consequences of failure aren't trivial. As neurobiologist Lou Reichardt, a Howard Hughes Medical Institute (HHMI) investigator at the University of California, San Francisco (UCSF), puts it, "The nervous system doesn't work unless it's wired properly."

The obvious significance of axon guidance has made understanding the system a very high priority for neurobiologists. But its complexity has made it a difficult problem to crack. Indeed, until the past few years, the maze was almost impenetrable. But a flood of recent publications from labs in Europe, Japan, and the United States shows that researchers are finally beginning to find their way through the axonal guidance labyrinth. The latest developments are reported in a remarkable flurry of eight papers appearing in the May and June issues of *Neuron* and in the 19 May issue of *Cell*.

This recent growth spurt started last year, when axonal guidance research got a big boost from the discovery by a group led by Marc Tessier-Lavigne, also an HHMI investigator at UCSF, of netrin-1. This protein is secreted by neuronal target cells into the surrounding tissue, where it attracts the correct axons. Developmental neurobiologists had been looking for such a diffusible chemoattractant for decades, but this was the first molecule that seemed to fit the bill. The new flood of papers fills out the axonal guidance story by showing that secreted chemorepellents — proteins that tell growing axons to "stay away" — are every bit as important as those that say "come here."

Researchers believe that there may be many such diffusible proteins that repel the searching axons, complementing the many secreted attractive factors that they believe exist. So far they have found three proteins with chemorepellent activity. These include, ironically, netrin-1, which apparently serves a dual purpose in axonal guidance, along with proteins known as semaphorin II and semaphorin III (or collapsin), which belong to a different family. And they have found that chemorepellent molecules appear to be secreted throughout the developing nervous system, occurring in the brain as well as the spinal cord and influencing pathfinding by several different types of neurons.

Adrian Pini of University College London, whose research 2 years ago provided the first evidence for diffusible chemorepellents, says that these latest discoveries "round off the possibilities for axonal guidance." He was referring to the fact that neurobiologists believe that there are four possible types of signals that could guide axons toward their targets: short-range repulsive and attractive cues provided by molecules on nerve cell surfaces, plus the diffusible chemoattractants and chemorepellents, which can act over longer distances.

The cell adhesion molecules involved in attraction were the first to come on the axonal guidance scene. In the early 1980s, researchers found that nerve cells carry molecules on their surfaces that can cause them to adhere to some other cells when they come into contact. The first evidence for short-range repulsion came later, toward the end of the 1980s, when three groups provided evidence that embryonic tissues carry membrane proteins that cause nerve cells to back away on contact. At the time, this finding was something of a surprise, according to axonal guidance researcher Corey Goodman, an HHMI investigator at the University of California, Berkeley. The reason: neurobiologists were "looking for 'turn-ons', not 'turn-offs,' " he says.

Researchers soon succeeded in identifying some of the cell surface molecules involved in both short-range attraction and repulsion, and these discoveries focused attention on contact-mediated cues for guiding axons to their targets. "Molecules involved in guidance were thought to be acting through adhesion," says Jonathan Raper of the University of Pennsylvania School of Medicine, whose team was one of those that discovered that neurons may be repelled when they contact another cell.

Despite this emphasis on the short-range, contact-mediated cues, however, there was also reason to believe that axons might be guided by signals that act over much longer distances as well. Indeed, the suggestion that neuronal targets might release such guidance molecules dates all the way back to work performed a century ago by Spanish neurobiologist Santiago Ramón y Cajal. Finding these molecules was another matter, however, and neurobiologists didn't get their hands on a good candidate until last year, when Tessier-Lavigne's group came up with netrin-1.

Their work showed that netrin-1 is made by the floor plate, a strip of tissue that runs along the lower (ventral) surface of the developing spinal cord, where it helps guide the axons of a particular type of spinal cord neuron: the commissural neurons, which relay information about pain and temperature to the brain. While these neurons are born in the dorsal (upper) surface of the spinal cord, their axons don't grow directly to the brain. They first project down to the ventral surface of the cord and cross the floor plate. Tessier-Lavigne's group showed that netrin-1 may help guide them along their way by acting as a diffusible positive signal to draw the commissural nerve axons to the floor plate.

One more to go

With this discovery, researchers had bagged examples of three of the four possible types of axonal guidance cues. Still missing were diffusible chemorepellents. But those wouldn't be long in coming. Pini had provided evidence for their existence in 1993, in experiments Tessier-Lavigne describes as "seminal work." In that research Pini exploited a culture system devised by Andrew Lumsden of Guy's Hospital in London and Alan Davies of St. Andrews University in Scotland that had already shown its merits by demonstrating the existence of diffusible attractants. The system involves culturing two samples of nerve tissue some distance apart within a collagen gel. Because the tissue samples aren't in contact, the only way they can influence each other is by releasing a substance that diffuses through the gel. In the experiments of Pini and his collaborators, axons of some neurons actually turned and grew away from other neurons, indicating that the others were secreting repellents.

At the time, the specific identities of the repellents weren't known. But even as the Tessier-Lavigne group was showing that netrin-1 is a chemoattractant for commissural neurons, there were also hints that this molecule might be a chemorepellent for other nerve cells. The first hint came from analysis of netrin-1's sequence, which revealed that it is the chicken equivalent of a protein from the nematode *Caenorhabditis elegans*, UNC-6, whose gene had been cloned by Ed Hedgecock of Johns Hopkins University, Joseph Culotti of Mount Sinai Hospital in Toronto, and their colleagues.

Studies of neuronal growth during development in *unc*-6 mutants indicate that the protein is involved in guiding axons of two classes of neurons that migrate in opposite directions. This suggested that UNC-6 attracts the one while repelling the other. The DNA sequence similarity suggested that the same might be true for netrin-1, and Tessier-Lavigne and UCSF colleague Sophia Colamarino set out to find out if that was in fact the case.

In the 19 May issue of *Cell*, they provide the answer — which is a resounding yes. For these experiments, the researchers chose embryonic, trochlear motor neurons, which help control eye movements. Unlike the axons of commissural neurons, trochlear neuron axons grow away from the floor plate, not toward it, a growth pattern that suggests the floor plate might repel the axons of the trochlear neurons. And that is exactly what Colamarino and Tessier-Lavigne found when they put trochlear neurons into the gel culture assay with floor plate tissue. To prove that the effect was due to netrin-1, the researchers transferred the netrin-1 gene into a type of non-neuronal cell that otherwise has no effect on axon guidance. The altered cells also repelled trochlear neuron axons. "The bifunctionality [of netrin-1] mirrors the dual action of UNC-6," Tessier-Lavigne says. "In one fell swoop, we've identified a diffusible attractant and diffusible repellent."

Netrin-1's dual guidance effects may not be limited to commissural and trochlear neurons. In the May and June issues of *Neuron*, a team led by Fujio Murakami of Osaka University in Japan describes results showing that axons of brain neurons respond to floor plate tissue much as spinal neurons do: Those that normally cross the floor plate during development are attracted by it, while those that don't cross it are repelled. "My guess is that many neurons share the [same guidance] mechanism," Murakami says. Further support for that idea comes from Sarah Guthrie of Guy's Hospital and Pini, who also report in the June issue of *Neuron* that the floor plate repels axons from certain motor neurons in the brain and spinal cord.

The Osaka team found that netrin-1 attracts the same neurons attracted by the floor plate in their studies, but haven't yet examined the protein's effects on the neurons that are repulsed. And while the London groups haven't done molecular studies to try to find out what was causing the effect they saw, the investigators predict that netrin-1 will be involved. "We haven't yet tested netrin-1 on the neurons, but we would be pretty surprised if it doesn't work," Guthrie says.

And netrin-1 isn't he only diffusible molecule that has a repelling effect on other axons. Semaphorin III/collapsin also plays that role, at least in culture. This protein has two names because it was discovered independently by two groups. Berkeley's Goodman and his colleagues originally discovered the first member of this family, semaphorin I, in the grasshopper as a cell surface protein that influences axon steering and inhibits axon branching. They then went on to identify genes for related proteins in the fruit fly and human. Two of these, semaphorins II and III, turned out to be secreted molecules, suggesting that they might be diffusible guidance molecules, although how they might function was something of a mystery.

Meanwhile, Pennsylvania's Raper and his colleagues had noticed that the growing tips of some axons from chickens collapse when they contact certain others in culture. Their assumption: The collapse of the growth cones, as the axonal tips are called, could well have been caused by some inhibitory substance on the axon they encountered. (This in fact was one of the early experiments indicating the existence of chemorepellent substances for neurons.) The Raper group set out to identify the molecule that causes the growth cone collapse, which they called "collapsin." They ultimately cloned and sequenced the gene in 1993, and its sequence revealed that collapsin is the chicken equivalent of semaphorin III, whose gene had just been cloned by the Goodman group.

In search of a role

While this work suggested that semaphorin III is a chemorepellent, the protein's exact role has been unclear. But another paper in the May issue of *Neuron* provides some answers. This work, a joint effort by the groups of Berkeley's Goodman, UCSF's Tessier-Lavigne, and Carla Shatz, also an HHMI investigator at Berkeley, shows that the protein specifically repels growing axons of one group of sensory neurons in the rat without affecting those of another. What's interesting, says Lumsden, is that the results are "thoroughly consistent" with the behavior of those neurons in the living animal.

The axons of both sets of neurons studied by the California team enter the spinal cord through the dorsal surface. After entering the cord, the axons of one set (which transmit sensory information about muscle stretch and position) proceed to the ventral region. The axons of the other group (which transmit pain and temperature information) stop in the upper dorsal region. Using the standard collagen gel culture assay, the three-way collaboration showed that semaphorin III repels the axons of this latter group but not those that normally penetrate the ventral cord. They also found that semaphorin III is made in the ventral half of the cord, putting it in just the right place to act as a "sieve" that can keep out pain and temperature neurons while letting the others in.

There is still a caveat about the roles proposed for both semaphorin III and netrin-1, however. Most of the work so far has been done with cultured neurons, and the ultimate proof of what the proteins do awaits the creation of animals in which the genes have been knocked out to see whether that produces the expected neuronal guidance defects.

In addition to the collaborative work on semaphorin III, evidence is piling up for other diffusible chemorepellents. The Goodman group, in a paper in the 19 May issue of *Cell*, presents data showing that semaphorin II has inhibitory effects on growing axons, although it apparently acts by a mechanism different from that of semaphorin III.

In this case, the evidence does come from studies in a living organism. Goodman and his colleagues genetically engineered fruit flies so that semaphorin II would be made in a muscle that does not normally express the protein. They found that this change blocked the ability of certain neurons to form synapses, the specialized connections between neurons and their targets, with the muscle. This suggests that semaphorin II acts much closer in than semaphorin III, which repels neurons before they get near the wrong place.

But these differences may only scratch the surface of the possible activities of semaphorin family members. For one thing, the family growing by leaps and bounds. In the May issue of *Neuron*, Heinrich Betz and his colleagues at the Max-Planck-Institut für Hirnforschung in Frankfurt, Germany, report cloning the genes for four new mouse semaphorins, while Raper's team has cloned four new genes from the chicken. (The paper will be in the June issue of *Neuron*.) A search of the databases indicates that humans also have several semaphorins, Goodman says.

Although the researchers do not yet know what the new semaphorins do, they have some clues. The Betz and Raper teams have found that each semaphorin gene has a specific expression pattern, with different genes being turned on in different tissues. This suggests that each repellent helps guide a different set of neurons. "As soon as you see there is a big family [of semaphorins], you can build in a great deal of specificity with just these repellent molecules," Goodman says. Indeed, agrees Lumsden, discovery of this family of molecules "suggests this is going to be a major factor in governing the projection patterns of neurons."

What's more, because some of the semaphorin genes are expressed in tissues, such as the lung, and others may function in the immune system, both places where they might be expected not to function in neuronal guidance, the importance of the proteins may extend beyond the nervous system. "We are all focusing on growth cone guidance, but they could be functioning in a variety of other things," Goodman speculates.

As the work progresses, researchers will try to decipher which neurons respond to which semaphorins, netrins, and other diffusible neuronal guidance molecules. They also want to know whether any of the semaphorins do double duty as netrin-1 does. Having such a system of diffusible guidance cues that repel, as well as attract, could be very useful to the developing nervous system, Pini suggests. If guidance works on such a "push-pull system," he says, "it would be a very efficient way of getting axons from one place to another over long distances."

An equally important question concerns the nature of the machinery by which neurons respond to the guidance molecules. This is especially true, Goodman notes, for molecules such as netrin-1 that can have opposite effects on different types of neurons. The reason for the differences must lie in the receptors and other components of the response machinery, as studies by the Culotti and Hedgecock groups already suggest. They have evidence that a protein called UNC-5 may be the receptor through which UNC-6 exerts its repellent — but not its attractive — effects on neurons.

And then there's the issue of what this growing repertoire of axonal guidance cues might mean clinically. Although the researchers are currently interested in the basic biology of the system, there has always been hope that understanding the molecules that guide neurons to their destination might lead to better treatments for people with spinal cord injuries and other nerve damage. Attractive molecules might be used, for example, to help direct regenerating neurons to their destinations. But researchers now know they will have to take the chemorepellents into consideration, too.

But despite the many mysteries remaining, the work has at least put diffusible guidance cues on a firm footing. And as Raper points out, the large number of unresolved issues has its own appeal. The work is "very exciting," he says, "because it raises a whole plethora of questions it will probably take years to answer." And, like diffusible molecules, those questions are drawing researchers over long distances toward the answers that are their targets. ❑

****

Questions:

1. What are axons?
2. What are the four possible types of signals that could guide axons toward their target cells?
3. Where is netrin-1 made in the body?

Answers are at the end of the book.

Losing a limb is bad enough but amputees have to suffer long afterwards with the burning, cramping or pain in that phantom limb. The reality of phantom-limb pain is that an amputee perceives their amputated arm or leg is still there. Just why this occurs has been subject to intense investigation especially in the somatosensory cortex of the brain that formerly handled sensations from the severed limb. A new study suggests that substantial cortical reorganization takes place as a result of neighboring sensory areas encroaching on the region previously owned by the amputated limb. With the new brain-imaging techniques available, researchers should know in the near future just how dramatic these changes might be.

Brain Changes Linked to Phantom-limb Pain

by B. Bower

For more than 100 years, physicians have published accounts of people who perceive an amputated arm or leg as if it were still there. Many amputees feel burning, cramping, or shooting pains in these phantom limbs, often at specific anatomical points. Reasons for the existence of these eerie, sometimes excruciating sensations remain uncertain, and medical attempts to ease phantom-limb pain usually fail.

A new study suggests that the most severe pain of this type occurs in people in whom, after amputation, an area of the brain formerly handled sensations from the severed limb undergoes extensive reorganization. Large-scale remodeling of this strip of sensory tissue may somehow alter the neural circuits involved in pain, thus resulting in phantom-limb pain, propose Herta Flor, a psychologist at Humboldt University in Berlin, and her colleagues.

"This is the first evidence that there is a central nervous system correlate of phantom-limb pain," says study coauthor Edward Taub, a psychologist at the University of Alabama at Birmingham. "It's one big piece of the jigsaw puzzle in understanding this phenomenon, but it's not the whole story."

Flor's team recruited 12 men and 1 woman, ranging in age from 27 to 73, who had had an arm amputated at least a year before the study. The scientists measured magnetic responses in participant's brains to light pressure on the intact thumb and pinkie finger, as well as on the left and right sides of either the lower lip or the chin. The researchers then mapped areas of magnetic activity onto a reconstructed image of the somatosensory cortex, a nerve impulse center for various body parts.

Since the left and right hemispheres of the brain regulate opposite sides of the body, stimulating the fingers sparked magnetic responses only in the hemisphere opposite the intact limb. A mirror image of the finger activation sites was then projected onto the hemisphere opposite the amputated arm.

This technique provides a good estimate of the location of somatosensory sites for an absent limb, Taub argues. Prior magnetic imaging work by another research team had found little variation in the location of corresponding parts of the somatosensory cortex in the hemispheres of nonamputated individuals.

Cortical reorganization, evidenced by substantial encroachment of sensory areas for the face into regions previously reserved for the amputated fingers, was most pronounced in the eight participants who suffered from phantom-limb pain, Flor's group reports in the June 8 NATURE. The greater the cortical shift, the more phantom-limb pain amputees reported.

The new data support earlier studies that charted cortical reorganization in monkeys after researchers cut nerve impulses to the animals' limbs.

Nervous system damage may trigger a strengthening of connections between somatosensory cells as well as the formation of new ones, Taub suggests. In some cases, an imbalance of pain messages from other brain areas may occur, leading to phantom-limb pain, he theorizes.

Or phantom-limb pain may result from a remapping of somatosensory areas that accidentally infringes on nearby pain centers, contends Vilayanur S. Ramachandran, a neurologist at the University of California, San Diego. Ramachandran and his coworkers reported last year that marked somatosensory reorganization had occurred in two amputees.

Further research with the more precise brain-imaging methods now available is needed to confirm the substantial cortical changes reported in adult amputees, notes Tim P. Pons, a neuroscientist at Wake Forest University's Bowman Gray School of Medicine in Winston-Salem, N.C.

****

Questions:

1. What is phantom-limb pain?
2. What area of the brain serves as a nerve impulse center for various parts of the body?
3. What happens in cortical reorganization as a result of nervous system damage?

Answers are at the end of the book.

Most people would rather think of the baseball legend when they hear the name Lou Gehrig and not the dreadful motor-neuron disease. Lou Gehrig's disease, amyotrophic lateral sclerosis (ALS), is characterized by the degeneration of descending motor pathways in the body and was responsible for ending his baseball career. Possible causes for the ALS disease are defective enzymes, environmental neurotoxins and infective agents. But the how and the why of ALS still remain unknown. Some of these questions are being addressed at the molecular level using transgenic animal models whose genetic make-up has been altered to produce the disease. The latest developments show the accumulation of neurofilaments in the cells bodies and proximal axons which prevents delivery of cell mitochondria and other essential components to the axon. Without sufficient mitochondria, the axons lack the metabolic energy to function and ultimately die. Making progress in this area will depend on finding the culprit that disrupts the neuronal dynamics responsible for putting people at risk for ALS.

Interfering With the Runners

by Scott Brady

In 1939, the New York Yankees baseball player Lou Gehrig ended his career after starting 2,130 consecutive games over 15 years. The sequence, which remains a record, finished only when an athlete legendary for his endurance was inexorably brought down by the most common of motor-neuron diseases, amyotrophic lateral sclerosis (ALS), now more familiarly known as Lou Gehrig's disease.

Gehrig's record may finally be beaten this year. The same cannot be said of the disease that bears his name, but since 1992 our knowledge of it has advanced on several fronts. The latest development is to be found in the paper by Collard and colleagues[1] Using a transgenic mouse model of ALS, they make a strong case that the aberrant accumulation of neurofilaments in motor neurons, and consequent disruption of axonal transport, can be responsible for neuron death.

More than 70 neurological disorders, affecting 1 in 20,000 people, involve specific loss of motor neurons.[2] Some result from defective enzymes, environmental neurotoxins or infective agents, but the causes of most are unknown. ALS itself is typically of adult onset, and involves a progressive degeneration of descending motor pathways that is generally fatal within five years of diagnosis; 90 percent of cases are sporadic and of uncertain aetiology.

The most notable events of the past three years have been the provisional identification of four distinct mechanisms that lead to specific loss of motor neurons. First, in 1992, Rothstein *et al.*[3] reported that ALS patients have reduced capacity for glutamate transport in the central nervous system, implicating excitotoxicity in cell death. Prolonged activation of glutamate receptors produces an influx of Ca^{2+} and neuronal degeneration.[4] Consistent with excitotoxicity, chronic inhibition of glutamate uptake leads to motorneuron death *in vitro*[5] and there is evidence of loss of glial glutamate transporters in motor areas of ALS patients.[6] The mechanism of selective loss is unknown, but neurons in transport-deficient areas would be vulnerable to excitotoxic pathology.

Second, in the following year the Cu/Zn superoxide dismutase gene (*SOD-1*) was shown to be altered in some cases of familial ALS. Several autosomal dominant alleles leading to disease have been identified, but only a quarter of familial ALS cases are traceable to mutant *SOD-1* genes. Although this suggested that oxidative damage might be responsible for motor-neuron death,[4] transgenic mice expressing mutant *SOD-1* develop ALS-like pathology despite having near normal enzyme activity,[8,9] and more than 95 per cent of ALS patients have wild-type *SOD-1*. In all, *SOD-1* linked familial ALS seems to stem from some unknown cytotoxic gain-of-function.[8,9]

Concurrently, autoantibodies to Ca^{2+} channels were detected in ALS patients.[10] These antibodies recognize several types of Ca^{2+} channel, inhibiting currents through L-channels but increasing them through P-channels.[11] Such currents could lead to neuronal degeneration through mechanisms similar to excitotoxicity, and the antibodies were cytotoxic *in vitro.*[12] These results pointed to a third, autoimmune, component of ALS.

Finally, indications of a fourth mechanism came from development of ALS-like pathology in transgenic mice overexpressing normal and mutant neurofilament subunits.[13-15] Not all transgenic mice develop ALS-like pathology,[16] but motor neurons are preferentially affected in those that do. Neurofilamentous accumulations in cell bodies and proximal axons are a hallmark of ALS, so these transgenics implied that disruption of neurofilament transport might cause motor-neuron death.

A variety of pathologies with different underlying causes are also associated with disruptions in axonal neurofilaments. Among them are diabetic neuropathy, Charcot-Marie-Tooth disease, and the effects of selected neurotoxins. Some have an associated increase in neuronal degeneration, but others lead to impaired function without a substantial loss of neurons. In most cases, however, pathological changes in neurofilaments are most apparent in large motor neurons, consistent with the fact that neurofilament expression is highest in these cells.[13-15]

Given that such diverse aetiologies have such similar consequences, the first lesson may be that neurodegenerative diseases are clinically defined by loss of specific neuronal populations, not by underlying mechanisms. Neurons are large, complex cells with unique interdependence on their targets and partners. Protein synthesis is restricted to cell bodies, and the metabolic needs of maintaining axons that, in humans, may be more than a metre long, place special demands on neurons. Both anterograde and retrograde degeneration can result from disruption of trophic relationships. Motor neurons affected by ALS are among the largest in the body, and extension of their axons into the periphery makes them especially vulnerable.

The observations of Collard *et el.*[1] provide a second lesson and illuminate the various mechanisms that lead to motor-neuron disease. In a transgenic mouse model of ALS produced by overexpressing the gene for human neurofilament heavy-subunit, the amount of material transported into axons is reduced, consistent with the distal atrophy seen in ALS axons. In particular, the number of mitochondria detectable in axons is reduced, meaning that axons may degenerate because they lack the metabolic energy to maintain even minimal function.

The implication is that accumulations of neurofilaments seen in ALS impede normal movements of other cellular structures, and that degeneration begins when supplies of new organelles and structures fall below a certain threshold. Other animal models of ALS also exhibit disrupted transport processes. In both glutamate excitotoxicity models[5] and the *SOD-1* transgenic mice,[8,17] the cell bodies of motor neurons display vacuoles and degenerating mitochondria. Similarly, autoantibodies to Ca^{2+} channels produce vesicle accumulations in the distal axons of motor neurons,[18] possibly reflecting disruption of trophic-factor delivery by retrograde transport. Finally, the Golgi apparatus is often fragmented in ALS motor neurons.[19]

Making progress on ALS will depend on getting to grips with its underlying complexity. Not every pathogenic mechanism producing motor-neuron death may disrupt axonal transport, and each disruptions may be secondary to the primary lesion in others. Nonetheless the observations of Collar *el al.* show that disrupting neuronal dynamics can lead to degeneration of specific neuronal populations. The point of disruption may vary with aetiology (mitochondria or Golgi, fast or slow transport, anterograde or retrograde movements, and so on), but it is the size and location of motor neurons that puts them at risk. ❑

References

1. Collard, J.-F., Cote, F., & Julien, J.-P., *Nature* **375**, 61-64 (1995).
2. Rowland, L.P. in *Neurodegenerative Diseases* (ed. Caine D.B.) 507-521 (Saunders, Philadelphia, 1994).
3. Rothstein, J.D., Martin, L.J. & Kunci, R.W. *New Engl. J. Med.* **326**, 1464-1468 (1992).
4. Coyle, J.T. & Puttfarcken, P. *Science* **262**, 689-695 (1993).

5. Rothstein, J.D., Jin, L., Dykes-Hoberg, M. & Kunci, R.W. *Proc. Natn. Acad. Sci. U.S.A.* **90**, 6591-6596 (1993).
6. Rothstein, J.D., Kammen, M.V., Levey, A. I., Martin, L. & Kunci, R.W. *Ann. Neurol.* (in the press).
7. Rosen, D.R. *et al. Nature* **362**, 59-62 (1993).
8. Ripps, M.E. *et al. Proc natn. Acad. Sci. U.S.A.* **92**, 689-693 (1995).
9. Gurney, M.E. *et al. Science* **264**, 1772-1775 (1994).
10. Appel, S. H., Smith, R.G., Engelhardt, J.I. & Stefani, E. *J. Neurol. Sci.* **118**, 169-174 (1993).
11. Llinàs, R. *et al. Proc. Natn. Acad. Sci. U.S.A.* **90**, 11743-11747 (1993).
12. Smith, R.G. *et al. Proc. Natn. Acad. Sci. U.S.A.* **91**, 3393-3397 (1994).
13. Cote, F., Collard, J.-F. & Julien, J.-P., *Cell* **73**, 35-46 (1993).
14. Xu, Z.-S., Cork, L.C., Griffin, J.W. & Cleveland, D.W., *Cell* **73**, 23-34 (1993).
15. Lee, M.K. & Cleveland, D.W., *Neuron* **13,** 975-988 (1994).
16. Eyer, J. & Peterson, A. *Neuron* **12**, 389-416 (1994).
17. DalCanto, M.C. & Gurney, M.E. *Am.J. Path.* **145**, 1271-1279 (1994).
18. Engelhardt, J.I. *et al. Synapse* (in the press).
19. Gonatas, N.K. *et al. Am. J. Path.* **140**, 731-737 (1992).

****

Questions:

1. What is amyotrophic lateral sclerosis?
2. What new DNA technology are scientists using to study ALS?
3. What mechanism can lead to motor-neuron death in ALS?

Answers are at the end of the book.

You just don't see anything in that three-dimensional psychedelic image no matter how long you stare at it. A new craze and long-time neurological mystery, autostereograms, are revealing some very interesting information to scientists about the way the brain works. While looking at such imagery, the brain compares the visual signals from both eyes since they each see an object but at slightly different angles. This provides the brain with information about the depth perception of the object. Recognition of the object comes later. To do this, the brain may be bouncing signals from the object-recognition and depth-perception regions back and forth to each other, which might explain the strange sensation people feel as they stare at the autostereogram and finally see the three-dimensional image.

Wallpaper for the Mind

by Carl Zimmer

In 1994, America became addicted to autostereograms—those swatches of psychedelic wallpaper that dissolve into three-dimensional images when you stare at the cross-eyed long enough. What the slack-jawed millions may not have realized, though, as they stared at books and posters, is that they were experiencing an enduring mystery of neurology: When the brain perceives a 3-D object, which comes first, the object or the 3-D?

The mystery has its roots in a previous 3-D craze, back in the nineteenth century. Victorian researchers discovered that they could create a 3-D illusion if they took photographs of an object with two cameras a few inches apart and had a person look at the images through a stereoscope, which allows each eye to see just one of the photos. The first stereograms were both a commercial smash and a neurological breakthrough: scientists realized that depth perception arises from the way the brain compares signals from the two eyes, which see an object from slightly different angles.

For over a century researchers assumed that the brain needed to recognize the signals as an object before it could compute the object's 3-D shape. In 1960, however, Bela Julesz, a psychologist now at Rutgers University, challenged that idea with a new kind of stereogram made of two identical fields of randomly scattered dots. In each field he drew an imaginary square around some of the dots and shifted them slightly to one side, filling in the blank gaps with more dots. If you looked at either field alone, you couldn't see the square. But when Julesz put both of them into a stereoscope, people saw a dot-covered square floating in front of a similar background. He and others concluded that depth perception is one of the first things the brain extracts from the visual signal, by comparing the left-eye and right-eye images dot by dot. Object recognition must come later.

Autostereograms, which can be viewed without a stereoscope, were invented in 1979 by psychologist Christopher Tyler. They consist of repeating vertical strips, like wallpaper, in which the pattern elements—random dots or something more complicated—have been shifted to one side, à la Julesz. They've been shifted in such a way that when you look at the pattern crossed-eyed (or in some cases look "through" it), the neighboring strips overlap. The pattern elements then fuse into left-eye and right-eye images of a single hidden object, which appears to be floating in space.

Although Bela Julesz can be considered the grandfather of the autostereogram craze, it turns out he wasn't entirely right about how the brain perceives 3-D objects. Vilayanur Ramachandran of the University of California at San Diego has shown as much by making a stereogram out of an optical illusion. The illusion consisted of three circles, each of which had a wedge cut out of it, arranged so the wedges formed the corners of an imaginary triangle. Even though there are gaps in the sides of the triangle, you see it whole because your brain fills in the gaps. Ramachandran created left-eye and right-eye versions of this illusion and had people look at them through a stereoscope. The people saw the illusory triangle floating in 3-D—even though the gaps prevented them from making the point-by-point comparison of left and right images that Julesz thought was essential to depth perception.

Ramachandran thinks the object-recognition and depth-perception regions of the brain may work in tandem, bouncing signals back and forth. That might explain the sensation some people have when looking at autostereograms: when they start to see the outline of a hidden object, the 3-D illusion suddenly kicks in. "The regions may be like two drunks," says Ramachandran. "Neither one can make it down the street alone, but if they lean up against each other, they stay upright." ❑

****

Questions:

1. Why are autostereograms a neurological breakthrough?
2. When the brain perceives a three-dimensional object, which comes first, the object or the 3-D?
3. How do the regions of the brain work together to process the information in an image?

Answers are at the end of the book.

Neuroscientists are claiming the dawn of a new era in dendrite biology. Advanced techniques are revealing that dendrites have a function beyond the computing of synaptic potentials that travel across its path. Recent results indicate that signals travel a two-way highway conveying messages to the cell body and then relaying them back to the outer reaches once again. And the signals are not limited to synaptic potentials alone for now there is evidence that action potentials are being initiated within the dendrites themselves. In addition, the action potential can travel back to the dendrite from the axon, thereby producing a better response to future signals. Called plasticity, this back-propagation of action potentials does strengthen the synapse and demonstrate that dendrites play an important role in actively shaping the responses of neurons and in the formation of learning and memory.

Dendrites Shed Their Dull Image

New techniques are revealing that dendrites, once thought to be mere adding machines, seem to be more actively involved in shaping the responses of neurons

by Marcia Barinaga

What goes on behind drawn curtains in the tight community of the brain? Consider the lacy, branching appendages of nerve cells known as dendrites. Dendrites have long had a reputation as solid but dull citizens of the neural metropolis: bean counters who passively add up the information they receive through the synapses dotting their surface. But it isn't easy to study dendrites directly, and some researchers have suspected that, like what goes on in the banker's basement at midnight, the secret lives of dendrites are more complex and interesting than their public image suggests.

Recent results indicate that those suspicions may be right. In the past few years, neurophysiologists, aided by new techniques that allow direct examination of dendrites, have begun to penetrate the mystery, producing some of the clearest insights yet into how dendrites work. Their studies mark the beginning of a "new era in dendritic biology," according to Columbia University neuroscientist Eric Kandel.

In that new era, researchers are confirming long-held suspicions that dendrites, far from being bean counters, play an active role in shaping the life of the neuron. Their finely branched network acts as a two-way highway, not only conveying incoming messages to the cell body, but also relaying information from the cell body back to their own outer reaches — information that may modify their responses to future signals. This ability of the dendrites to react and adjust their activity is likely to play an important role in such mental processes as learning and memory. The work has "really put the spotlight on the dendrites," says neuroscientist and neural modeler Terrence Sejnowski of the Salk Institute.

The "bean-counter" image of dendrites stems from the fact that their job as information receivers requires that they pass along to the cell body the synaptic potentials, the little dollops of charge that enter the dendrites when neurotransmitter molecules activate synapses. During the journey to the cell body, each synaptic potential adds up with all the others moving through the dendrites. If their sum is large enough when they reach the cell body, they trigger an action potential by causing sodium ions to flow into the cell. The action potential then sweeps down the axon, the part of the nerve cell responsible for carrying the signal to other neurons.

In the traditional dogma of neuroscience, action potentials were limited to axons, and synaptic potentials to dendrites. "This made a clear distinction between the input and the output of the cell," says Salk Institute neurophysiologist Chuck Stevens. But more than 30 years ago, cracks began to appear in that simple model, as researchers reported evidence of action potentials in the dendrites. Some theorists were

Reprinted with permission from *Science*, April 14, 1995, Vol. 268, pp. 200-201.

intrigued by this possibility and began to consider what action potentials might be doing in the dendrites. Others pooh-poohed the idea, saying that action potentials sweeping through the dendrites would foul up their ability to add up synaptic potentials.

The debate was complicated by the fact that dendrites were considered too fine to be impaled easily with electrodes. Their diameter is a mere 1/1000th the diameter of the squid giant axon, which was used for studies of action potentials in axons, and 1/10th the diameter of a typical neuron cell body, which is where neurophysiologists generally put their electrodes. That meant that, at first, all the evidence for action potentials in the dendrites was gathered indirectly. In the 1970s, however, Rodolfo Llinás's team at New York University poked electrodes directly into dendrites of neurons from the cerebellum and recorded action potentials there. But critics worried that the delicate dendrites may have been damaged by the impaling electrode, producing an erroneous result.

Even those researchers who believed the action potentials were real couldn't agree on which way they were going. Some experiments suggested action potentials may initiate within the dendrites themselves. Others suggested they started in the axon and spread into the dendrites from there. The field was clearly in need of new approaches.

A key advance came last year from Greg Stuart, a postdoc with Bert Sakmann at the Max Planck Institute for Medical Research in Heidelberg, Germany. Sakmann shared the 1991 Nobel Prize for developing a technique called patch-clamping, which replaced sharp, impaling electrodes with smooth-ended electrodes which are pressed up against the neuron's outer membrane to form a tight electrical seal. Stuart had developed methods for patch-clamping dendrites that took advantage of new contrast-enhancing microscopy techniques to bring the dendrites sharply into view.

Using the new method, Stuart put separate patch clamps on the cell body and dendrites of individual neurons in slices of cerebral cortex from rat brains. His results, reported in the 6 January 1994 issue of *Nature,* showed that, when the dendrites were stimulated and the neuron fired, the resulting action potential was registered first by the electrode on the cell body and then, milliseconds later, by the electrode on the dendrites.

This was the first direct evidence that action potentials travel into the dendrites after being triggered in the axon. "This dual-patch technique that the Sakmann lab introduced really nailed this subject down in a way that wasn't capable of being done in the past," says neuroscientist William Ross of New York Medical College in Valhalla.

The idea of "back-propagating" action potentials (so called because they travel against the main flow of information from dendrite to cell body to axon) had been kicked around among theorists for years. It has a certain appeal, because it could serve as a feedback signal to the dendrites and the synapses on their surfaces.

Researchers have long been known that neurons can adjust the strength of their response to incoming signals, a process called plasticity that is thought to be important for learning. For example, in one form of plasticity, synapses that are active when a neuron fires become stronger, so that they give a bigger response to future signals. Back-traveling action potentials seem "like the ideal way to tell the synapses something happened," says neuroscientist Dan Johnston of Baylor College of Medicine in Houston. "How else will they know the cell fired?"

To find out whether the action potentials might in fact have such a role, Sakmann postdoc Nelson Spruston turned to the neurons in which synaptic plasticity has been most intensely studied, the CA1 neurons of the hippocampus, a brain structure involved in some forms of learning. Sakmann's group reports that back-propagation occurs in the CA1 neurons, although they don't yet know if it influences plasticity.

Although they have not answered that question, they made an unexpected discovery that might relate to plasticity. They noticed that sometimes, after a neuron had fired several times in succession, the electrode on the dendrite would register a much-reduced signal, as if the later action potentials in the series fizzled out before reaching it.

Clarification of what was

happening came from Sakmann postdoc Yitzhak Schiller, who was studying action potentials by injecting neurons with calcium-sensitive dyes. Action potentials let calcium into the cell, so Schiller could use the dyes to record the path of the action potentials out to the finest tips of the dendrites. He found that when the neuron fired off a series of action potentials in a row, the first one would spread unhindered throughout the dendrites, but subsequent ones would not make it past some of the dendrites' branch points.

When Schiller and Spruston saw each other's results, "we got really excited," Spruston says. "It all fit together very nicely. ... You have the idea of all these branch points acting like switches that can control the number of action potentials that get through those points." Their finding was confirmed by Joseph Callaway, a postdoc in Ross's lab at Valhalla, who has similar results with calcium-sensitive dyes submitted for publication.

Still, a big question remains unanswered: What is the significance of this switching? If the back-propagating action potentials do contribute to plasticity, says Columbia's Kandel, the fact that branch points can act as switch points means that "one [branch] will [receive] the action potential and one will not. ... You will have the capability in one branch for plasticity that the other branch will not have." That adds to the computing potential of the dendrites, as selective strengthening of some branches would give a boost to signals arising there, leaving other branches unaffected.

Despite their apparent usefulness, backward-traveling action potentials are not universal in dendrites. NYU's Llinás found, in neurons called Purkinje cells, that action potentials begin with the dendrites themselves and travel toward the cell body. But these are not typical action potentials. Purkinje cells lack sodium channels in their dendrites, and the action potentials Llinás observed are carried by calcium ions instead. Indeed, the lack of sodium channels in Purkinje cell dendrites may explain why action potentials that start in the cell bodies of those neurons do not travel backward into the dendrites, a fact that Stuart and Michael Häusser confirmed with dual patch-clamp recording, as reported in a paper in last September's issue of *Neuron*. While the role of the calcium-based action potentials in Purkinje cells is not known, their existence attests to a different "philosophy of [function] of the two cell types," says Llinás.

Besides resolving the question of whether dendrites carry action potentials, recent research is clearing up another long-standing mystery about dendrite action, and that is how synaptic potentials travel through the dendrites. Action potentials travel "actively," which means that, as they pass through the neuron, they open ion channels, allowing positively charged ions to flow into the cell and add their charge to the traveling signal, like springs continually renewing a river's flow. Synaptic potentials, on the other hand, were thought to travel passively, like creeks in dry country, simply flowing along the inside of the membrane with no additional inputs.

The problem is that a potential traveling in the manner loses charge at it goes, and so synaptic potentials from the most distant dendrites would virtually disappear before reaching the cell body. But researchers long ago showed that those distant synapses are able to fire a neuron, which means those synaptic potentials must be able to make the long journey. "You wonder how could these events on the distal dendrites ever make it," says Yale University neurobiologist Tom Brown. "The obvious answer is there has to be some booster out there."

Studies with ion-sensitive dyes had suggested that there are ion channels in the membranes of dendrites that might act as boosters. Baylor's Johnston and postdoc Jeffrey Magee provide more direct evidence that that is the case. They patch-clamped the dendrites and studied individual channels in the patches of membrane beneath the electrode to find out under what conditions the channels opened. "We allowed the cell to experience normal synaptic potentials," says Johnston, "and asked the question of whether or not, in the little patch of membrane, the channels will open during the normal physiological event." They found that voltage-dependent sodium and calcium channels in the membrane open not only in response to action potentials, but also when synaptic potentials pass by on their way to the cell body. "It's like power lines," says Johnston. "You have transformers and amplifiers along the way" to maintain and modify the signal.

Johnston points out that the entry of calcium ions may also be an important cause of plastic changes in the dendrites. Indeed, calcium is a powerful intracellular signal. Whether let in by back-propagating action potentials or by synaptic potentials, it is bound to play a role in shaping the future response of the dendrites. "There are likely to be many, many processes triggered by calcium," says neuroscientist Roberto Malinow of Cold Springs Harbor Laboratory, "synaptic potentiation, ... regulation of ion channels, and even [dendrite] growth."

Despite all the recent revelations, neuroscientists have a long way to go before they understand how all this activity in the dendrites contributes to mental processes. But the next round of experiments addressing that question is already well under way in many labs. For example, Sakmann's group is testing to see whether back-propagating action potentials are necessary to strengthen synapses, and Sejnowski's team at Salk has a paper in press at the *Journal of Physiology* in which they report that the firing of action potentials in CA1 cells strengthens the cells' response to later signals, possibly through bigger boosts to the synaptic potentials as they travel to the cell body.

Ultimately the same techniques will be applied to many other neuron types and will reveal even more variety in the activities of dendrites. "It is all coming together very rapidly," says Sejnowski. And as it does, the staid image of the dendrite is being replaced by something far more intriguing. ❑

****

Questions:

1. What happens to synaptic potentials that the originate in the dendrites?

2. What is the patch-clamp technique?

3. What value do backward-traveling action potentials have in the dendrite?

Answers are at the end of the book.

In order to sustain a high level of neurotransmitter release at the nerve synapse, an army of synaptic vessels must be waiting in reserve. Since the demands for these synaptic vessels can be great during high-frequency nervous stimulation, the process of exocytosis must become a regulated process in order to remain efficient. Researchers have uncovered a protein concentrated in nerve terminals within the reserve pool of vesicles which appears to have a dual regulatory function at the synapse. Working behind the scenes is a protein called synapsin which serves as a link between the vesicles and the cytomatrix of the nerve terminal, helping to recruit and control the availability of synaptic vesicles for release. The role of synapsin in synaptic physiology offers new insight into how the nervous system is able to carry out its complex and dynamic functions even under the most demanding circumstances.

Keeping Synapses up to Speed

by Pietro De Camilli

The proper functioning of the nervous system demands that synapses should operate efficiently over a wide range of frequency stimulation. To guarantee neurotransmitter secretion during bursts of intense activity, nerve terminals store large pools of synaptic vesicles in reserve. Two reports published[1,2] by Rosahl *et al,* and Pieribone *et al,* provide new insight into the mechanisms involved in the build-up of these reserves and reveal the consequence to synaptic physiology when they become depleted.

Depolarization of a nerve terminal inducted by an action potential triggers exocytosis from a pool of ready-to-fuse synaptic vesicles — on average 10-20 vesicles at central synapses — which are docked at active zones of the presynaptic plasmalemma[3]. Membranes of fused vesicles are then reused for the generation of new fusion-competent vesicles by membrane recycling. But because this process is slower than the speed at which synaptic vesicles can be depleted from active zones during intense activity[3,4], presynaptic compartments are in danger of becoming rapidly depleted of synaptic vesicles during high-frequency stimulation unless new fusion-competent vesicles can be drawn rapidly to active zones from a reserve pool.

Synaptic vesicles are a special form of the endosome-derived vesicular carriers that are present in all cells, so the basic mechanisms involved in exocytosis and recycling of synaptic vesicles are evolutionary conserved[5]. However, only in the case of synaptic vesicles is exocytosis tightly regulated, implying that neuron-specific mechanisms are in operation. The protein synapsin is an abundant nerve-terminal specific protein which is thought to play a part in some aspect of the secretion process that is unique to synaptic vesicles[6-8].

Synapsin is a collective name for two very similar proteins, synapsin I and synapsin II, which were first identified in the 1970s by Greengard and co-workers as endogenous brain substrates for cyclic AMP and CA^{2+}-dependent phosphorylation[8]. Synapsin I, but not synapsin II, contains a highly basic, proline-rich region at its carboxy terminus[7]. Synapsin is concentrated in nerve terminals, where it is peripherally associated with the cytoplasmic surface of synaptic vesicles. *In vitro* studies have shown that purified synapsin binds actin with high affinity, that it can crosslink synaptic vesicles to actin, and that its interaction with both actin and vesicles can be modulated by phosphorylation[8,9]. Manipulations of nerve terminals that enhance neurotransmitter secretion also induce an increase in the state of phosphorylation of synapsin. This change is paralleled by an increase in the amount of synapsin recovered in soluble fractions, suggesting that, *in vivo*, the interaction of synapsin with vesicles is dynamic[8].

It has been proposed that synapsin provides a regulated link between synaptic vesicles and the cytomatrix of nerve terminals and that is may control the availability of synaptic vesicles for release. This would imply that the protein has a key role in both the ontogenesis and maintenance of synapses[8,10]. It was therefore a surprise to find that mice lacking synapsin I displayed a relatively mild phenotype, with no apparent change in a variety of functional and morphological parameters and only an altered paired-pulse facilitation (the phenomenon by which the second of two stimuli separated by a short interval elicits a greater release of neurotransmitter)[11]. Although additional deficits have come to light in more detailed investigations of these mice[1,2], the synapsin I knockout results show that the function of synapsin I is dispensable for the basic process of neurotransmitter release at the synapse. The interpretation of these findings, however, is hampered by a possible redundancy between the functions of synapsins I and II.

Rosahl *et al.* have now generated knockout mice lacking synapsin II, or both synapsin I and II, and performed a thorough anatomical, histological and functional characterization of these animals[1]. Even the double-knockout mice were viable, but they developed severe generalized seizures. Although the general architecture of the neuropile was unchanged, synaptic vesicle number and the level of intrinsic synaptic vesicle membrane proteins (but not of the corresponding messenger RNAs and of peripheral vesicle membrane proteins) were severely reduced. At hippocampal synapses, post-tetanic potentiation (an increased secretory response to a stimulus that follows a high-frequency train of action potentials) was drastically decreased and activity-dependent synaptic depression (the run-down of secretory response during repetitive stimulation) was increased. Synapsin II-knockouts had a similar but less severely damaged phenotype, and generally the severity of the phenotype was roughly related to the number of missing synapsin alleles[1].

In an independent study, Pieribone *et al.* have tested the effect of the acute antibody-medicated disruption of synapsin function on the *'en passant'* synapses of lamprey reticulospinal axons[2]. The anatomical features of these axons make them a very elegant system for this type of study. The injection of antibodies directed against the conserved carboxy terminus of mammalian synapsins Ia and IIa, but not injections of control antibodies, caused a virtually complete loss of synaptic vesicle clusters at synapses, with the exception of the pool of vesicles close to the presynaptic plasmalemma. This morphological change did not affect the level of transmitter release elicited by low-frequency stimulation, but produced a pronounced reduction in release after high-frequency stimulation[2].

The results obtained in the lamprey are strikingly complementary to the results from the knockout mice. In both cases, disruption of synapsin function equals fewer vesicles at the synapse and, as a consequence, inability to maintain high levels of release during a burst of synaptic activity. The more drastic phenotype observed in the antibody-injection studies and the different patterns of vesicle depletion (more global in the knockouts and very selective for the vesicle pool distal to active zones in the lamprey) may be at least partially explained by differences between the acute and the chronic disruption of a protein. In a chronic disruption, compensatory changes may come into play. In principle, the results of the two studies could be explained by defective recruitment of synaptic vesicles at presynaptic sites, or by impaired reuse, leading to increased synaptic vesicle catabolism[1]. As disappearance of the vesicle cluster in the lamprey study was facilitated by, but not dependent on stimulation of the synapse[2], Pieribone *et al.* take these data as strong support for the idea that synapsin is important for the recruitment of synaptic vesicle clusters at release sites.

A role of synapsin trapping synaptic vesicles into a presynaptic matrix is consistent with several previous studies[8,10] and with the mice knockout results. Inefficient retention of synaptic vesicles in nerve terminals may result in an increased turnover of intrinsic synaptic vesicle proteins. Peripheral membrane proteins that associate with synaptic vesicles only in nerve endings may be unchanged[1] because their accumulation at the axonal periphery is likely to be regulated by distinct mechanisms.

This putative role of synapsin is also in agreement with the absence of the protein from ribbon synapses of axonless cells of sensory organs[12]. These synapses, which use synaptic vesicles that are otherwise very similar morphologically and biochemically to *bona fide* synaptic vesicles[12], have unique functional properties that correlate with special structural features. Neurotransmitter release is regulated tonically by graded changes in membrane potential, and high rates of synaptic vesicle exocytosis can be sustained for prolonged periods[13-15]. At these synapses, synaptic vesicles are concentrated close to active zones by dense cytoplasmic bodies (which are ribbon-like in photoreceptor and bipolar cells of the retina) completely coated by synaptic vesicles. It was proposed that synaptic vesicle trapping mediated by the ribbon may guarantee a constant supply of vesicles during prolonged depolarizations[13,15]. The dense body of ribbon synapses and a synapsin I-containing matrix may have corresponding functions at ribbon synapses and at conventional axonal synapses respectively.

It must be emphasized, however, that even these new studies do not allow unequivocal interpretations of synapsin I function, and other scenarios are possible. For example, changes in paired-pulse facilitation (a phenomenon that occurs on a scale of milliseconds) observed in synapsin I but not in synapsin-II or double-knockout mice[1], may indicate a specific role for synapsin I very proximal to fusion. On the other hand, this effect may be indirect and mediated by proteins that normally interact with the proline-rich carboxy terminal region of synapsin I (for example SH3-containing proteins[16]). In the synapsin-deficient mice, the selective increase in Rab5, a protein thought to be involved in the endocytic pathway[17], cannot yet be explained.

Irrespective of the interpretation, the phenotype induced by synapsin disruption offers fresh insights into how synapses work and shows that a key function of synapsin is to ensure a steady supply of fusion-competent vesicles to release sites. Synapsin plays a fundamental role in conferring upon synapses the plasticity required for the complex functions of the nervous system. ❑

References

1. Rosahl, T. W. *et al. Nature* **375**, 488-493 (1995).
2. Pieribone, V. A. *et al. Nature* **375,** 493-497 (1995).
3. Stevens, C. F. & Tsujimoto, T. *Proc. Natn. Acad. Sci. U.S.A.* **92,** 846-849 (1995).
4. Ryan, T. A. & Smith, S. J. *Neuron*, **14,** 983-989 (1995).
5. Ferro-Novick, S. & Jahn, R. *Nature* **370**, 191-193 (1995).
6. DeCamilli, P., Harris, S. M., Huttner, W. B. & Greengard, P. *J. Cell Biol.* **96,** 1355-1373 (1983).
7. Sudhof, T. C. *et al. Science* **245,** 1474-1480 (1989).
8. Greengard, P., Valtorta, F., Czernik, A. J. & Benefenati, F. *Science* **259**, 780-785 (1993).
9. Bahler, M. & Greengard, P. *Nature* **326**, 704-707 (1987).
10. Han, H. Q. *et al. Nature* **349**, 697-700 (1991).
11. Rosahl, T. W. *et al. Cell* **75,** 661-670 (1993).
12. Mandell, J. W. *et al. Neuron* **5,** 19-33 (1990).
13. Parsons, T. D., Lenzi, D., Almers, W. & Roberts, W. M. *Neuron* **13,** 875-883 (1994).
14. VonGersdorff, H. & Matthews, G. *Nature* **367,** 735-739 (1994).
15. Rao-Mirotznik, R., Harkins, A. B., Buchsbaum, G. & Sterling, P. *Neuron* **14,** 561-569 (1995).
16. McPherson, P. S. *et al. Proc. Natn. Acad Sci., U.S.A.* **91,** 6486-6490 (1994).
17. Singer-Kruger, B. *et al. J. Cell Biol.* **125,** 283-298 (1994).

****

Questions:

1. What event initiates the process of exocytosis?
2. What are synaptic vesicles?
3. What role does synapsin play at the synapse?

Answers are at the end of the book.

It may be easy to dissect and understand how computers gather and store information, but have you ever wondered how the brain does this with memories? Some scientists are pointing their fingers at the putative retrograde messenger nitric oxide, spreading the effects of long term potentiation (LTP) to nearby synapses by diffusing back across the synapse to the presynaptic cell strengthening that specific synapse. According to their findings, it is involved in enhancing neural connections in localized regions of the brain and even meddling in neighboring pathways that are not directly connected. In terms of brain development and memory storage, how nitric oxide may be strengthening the bond between two or more neurons is a question that deserves further investigation. With the help of computers, researchers will be able to model the workings of the brain and arrive at some definitive answers concerning nitric oxide's role in spreading potentiation.

Learning by Diffusions: Nitric Oxide May Spread Memories

by Marcia Barinaga

For an esoteric phenomenon, long term potentiation (LTP), has received a lot of attention in recent years. That's because this process strengthens the synapses that provide the functional links between neurons. Consequently, it may be central to both brain development and memory storage. Part of LTP's appeal has been that it appeared to be exquisitely specific, able to pick out and strengthen one synapse from among the thousands that dot a particular neuron. To brain modelers, that specificity meant each synapse could function as an independent memory-storage unit — the biological equivalent of the computer bit.

But that view of the synapse may now be forced to change. Dan Madison of Stanford University School of Medicine and his former postdoc Erin Schuman, now an assistant professor at the California Institute of Technology, report findings suggesting that LTP is not as specific as had been thought. In fact, it can apparently be spread to synapses on neighboring neurons by a diffusible signal — and that signal may be none other than nitric oxide (NO), a highly reactive, soluble gas. Although this finding may shatter the former image of LTP, it fits what's known about NO, which has already been implicated as a diffusing signal in LTP.

Finding that potentiation can spread to nearby synapses provides "a phenomenal new insight into the nature of LTP," says Columbia University neuroscientist Eric Kandel. Says Roger Nicoll, a neuroscientist at the University of California, San Francisco: "This is without a doubt the most direct evidence that there is some transfer of potentiation through a diffusible message." And he adds that the finding is "very provocative. Nature has gone to elaborate lengths to create a structural edifice that can give you synapse specificity. To then just degrade the process and let it spread around a bit, makes it seem like Nature blew it somehow."

The first evidence for spreading potentiation came in 1989 from Tobias Bonhoeffer, Volker Staiger, and Ad Aertsen, at the Max Planck Institut für Biologische Kybernetik in Tübingen. LTP can be triggered in a single synapse by simultaneously stimulating both the "presynaptic" neuron, which sends a signal across the synapse, and the "postsynaptic" neuron, which receives the signal. When the Bonhoeffer group used this technique to induce LTP in a synapse in a slice of rat hippocampus (a brain region involved in some types of learning), they found that the responses at synapses on neighboring neurons appeared to be strengthened as well.

At the time, LTP was thought to be highly specific, and their result was so iconoclastic that it didn't win rapid acceptance. Then, 2 years later, independent evidence of spreading potentiation came up in work that Schuman and Madison were doing on the role of nitric oxide in LTP. NO had been implicated as the sought-after "retrograde signal," which is thought to be made by the postsynaptic neuron during LTP and to diffuse back across the synapse to the presynaptic cell, presumably strengthening that specific synapse. But Schuman and Madison found that if they blocked NO production in a specific synapse, that synapse could still be strengthened, provided that others nearby were undergoing LTP. It seemed that NO from those synapses could rescue the synapse whose NO production was blocked.

To explore this idea further, Schuman and Madison triggered LTP in single synapses in slices of rat hippocampus. Using more precise detection methods than Bonhoeffer's, they measured the effect on nearby synapses on neighboring neurons and found, as Bonhoeffer had, that the synapses were strengthened. Their painstaking methods and the sheer number of neurons they sampled in 3 years of work on the project won over many researchers who had been reluctant to buy the notion of spreading potentiation. "I was very skeptical initially," says neuroscientist Michael Stryker of the University of California, San Francisco, but technically the new work seems really sound."

Madison and Schuman went on to test the role of NO in spreading potentiation, and found that when they blocked production of the gas in the postsynaptic neuron, they also blocked the strengthening of neighboring synapses. But Madison cautions that these findings don't prove that NO is the signal; it might instead trigger formation of another molecule that does the actual signaling.

Whatever the signal that wafts out from a potentiated neuron, a large number of synapses on many other neurons will be within its reach. Schuman and Madison had no way of telling whether all or just some of those synapses would be strengthened. But data from another research group does address that question. Last June, Kandel and his colleagues Min Zhuo, Scott Small, and Robert Hawkins reported in *Science* that spritzing NO or carbon monoxide (another candidate for the diffusible message) onto hippocampal slices strengthened only those synapses that were already receiving nerve impulses.

If potentiation spreads only to active synapses, it would be ideally suited to the job of helping to refine nerve connections in the embryonic brain. The mature brain is organized around clusters of neurons that respond to similar stimuli: such as the columns of neurons in the visual cortex that all respond to input from the same eye. To achieve that organization, embryonic neurons start with excess connections, then prune back, selectively keeping those that receive input similar to that received by their neighbors. In 1990, Read Montague, Joseph Gally, and Gerald Edelman of the Neurosciences Institute in New York City predicted that a diffusible substance such as NO could help determine which neurons are kept by broadcasting the news that a potentiating signal had been received, and triggering other neurons to strengthen any synapses active at the same time. So far, however, there are only preliminary — and conflicting — reports on whether NO actually performs such a function in developing brains.

Another intriguing question is what spreading potentiation means for learning and memory. For starters, it means the individual synapse cannot be the "computer bit" of the brain. "Instead of thinking of a synapse as representing a piece of information, you can now begin thinking of a population of [potentiated] synapses" acting together, says Salk Institute neural modeler Terrence Sejnowski. And while that reduces the storage capacity of the brain, Sejnowski believes that spreading out the storage may confer unknown advantages. He and Montague, who is now at Baylor College of Medicine in Houston, are developing computer models to test that hunch. So it may turn out that Nature didn't make such a mistake after all. ❑

****

Questions:

1. What neural mechanism is thought to play a role in memory and learning?

2. What is nitric oxide?

3. What study did scientists perform to test the hypothesis that nitric oxide might evoke potentiation in one synapse of a neuron and in nearby synapses?

Answers are at the end of the book.

Nitric oxide is considered by many in science to be one of the most important messenger molecules in the body. A character cast in a multitude of different roles, the quest within the scientific community to discover the truth about nitric oxide has been fast-paced and exciting to say the least. Research has shown the illusive molecule to be toxic to tumor cells and bacteria and it allows neurotransmitters to dilate the smooth muscle of blood vessels. Perhaps even more important is the function of nitric oxide in nerve cells in the brain and in the peripheral autonomic nervous system. Tracking down this short-lived molecule isn't easy which is why scientists are using the enzyme nitric oxide synthase, the maker of nitric oxide, to follow its function in the body. The newest discovery is the existence of neuronal nitric oxide synthase in skeletal muscle. With nitric oxide present, it can mediate skeletal muscle relaxation, just as it does in the smooth muscle of blood vessels.

More Jobs for that Molecule

by Solomon H. Snyder

It is not so long ago biologists became interested in nitric oxide (NO), but in that short period it has been found to carry out more important functions than virtually any other known messenger molecule.[1,2] Kobzik *et al.* have found yet further tasks for this versatile marriage of nitrogen and oxygen. Their paper contains a sizable body of data detailing the effects of NO in rat skeletal muscle.

Biologically, NO was first characterized as endothelial-derived relaxing factor, being formed by endothelial cells in blood vessels and diffusing to the adjacent smooth muscle to cause vasodilatation. Endogenous formation of NO is now regarded as the main determinant of blood pressure, and the principal drugs used in treating angina act by being converted to NO. About the same time macrophages — white blood cells that engulf and kill tumour cells and bacteria — were shown to act by forming large amounts of NO when stimulated by endotoxin.

Subsequently, the molecule has been identified as a neurotransmitter in the brain and peripheral autonomic nervous system, where it mediates a large number of biological functions ranging from learning and memory to peristalsis and penile erection. These three principal sites of NO formation are paralleled by the existence of three distinct forms of NO synthase (NOS) derived from distinct genes.

Neuronal NOS, the first to be molecularly cloned, possesses numerous regulatory sites, including a critical binding site for calmodulin which enables NOS production to be included by stimulation of glutamate *N*-methyl-D-aspartate (NMDA) receptors that open calcium ion channels. Excess release of NO may be responsible for severe neurological damage in vascular stroke, as NOS inhibitors block NMDA neurotoxicity as well as stroke damage.[2] Moreover, mice in which the gene for neuronal NOS has been deleted manifest much less brain damage following vascular stroke than wild-type mice.[4] Endothelial and neuronal NOS enzymes are generally regarded as constitutive, with increases in enzyme activity being triggered by stimuli that increase calcium influx. By contrast, the macrophage enzyme is inducible[5] — in their resting state macrophages have no detectable NOS, but stimulation by endotoxin triggers a massive increase in synthesis of the protein.

But these are stereotypes. Following the initial characterization of the three main subtypes of NOS and NO functions, a second wave of NO research is rapidly overturning them. For instance, it has turned out that the 'non-inducible' neuronal NOS is in fact inducible. Following damage to peripheral or central neurons, high concentrations of neuronal NOS are detected in neurons that normally lack the enzyme.[6,7] During embryonic development, major transient expression of neuronal NOS occurs in a neuronal pathway proceeding from the cerebral cortex to the thalamus, in olfactory neurons and in sensory ganglia.[8]

Macrophage NOS is not expressed only in macrophages,

Reprinted by permission from the author, Solomon H. Snyder and *Nature,* Vol. 372, pp. 504-505.

but in many other tissues as well; indeed, it seems possible that an inducible macrophage-like NOS occurs in virtually all tissues as a primitive sort of immune system.[1] Endothelial NOS does not occur only in blood vessels. The enzyme is also concentrated in a variety of neuronal structures in the brain, especially the pyramidal cells of the hippocampus, where it may provide the NO postulated to serve as a retrograde messenger in long-term potentiation because these cells lack classical neuronal NOS.[9] Finally, neuronal NOS is not confined to neurons, in that is has been detected in the epithelium of the respiratory tract.[10]

Kobzik and associates[3] now add to the list. They describe the existence of neuronal NOS in skeletal muscle, where it is concentrated in fast fibres and seems to have a physiological role in relaxing muscle. Last year, neuronal NOS messenger RNA was identified in samples of human skeletal muscle by Nakane *et al.*[11], and Kobzik *et al.* have followed up on that result by employing immunohistochemical techniques to localize the enzyme primarily to type II fast-contracting muscle fibres. Enzyme activity and immunostaining are concentrated in the membrane fraction of muscle fibres, challenging another NO catechism that holds that neuronal and macrophage NOS are always cytoplasmic, while only endothelial NOS is associated with cell membranes.

What does skeletal muscle NOS do? Inhibitors of the enzyme increase skeletal muscle contraction, effects that are reversed by NO donors, indicating that NO presumably physiologically relaxes skeletal muscle in much the same way that it relaxes the smooth muscle of blood vessels.[3] So, how might it bring about this effect? In blood vessels NO binds to haem in the active site of guanylyl cyclase to augment its activity with the newly formed cyclic GMP, eliciting relaxation by some undefined mechanism. Evidence for some involvement of cGMP in NO activity in skeletal muscle comes from observations that 8-bromo cGMP elicits modest relaxation, as does an inhibitor of phosphodiesterase, whereas an inhibitor of guanylyl cyclase causes contraction.[3]

Kobzik *et al.*[3] suggest another intriguing way in which NO might regulate skeletal muscle. They observe release of NO but not of reactive oxygen intermediates in uncontracted skeletal muscle, whereas in contracting muscle they detect high levels of the reactive oxygen intermediates (presumably comprising superoxide and possibly other oxygen free radicals). It is well known that NO binds to superoxide, thus inactivating it. On the other hand, the combination of NO and superoxide can form peroxynitrite which decomposes to hydroxide free radical, one of the most reactive of all free radical species. One can speculate that NO modulates the effects of oxygen free-radical species formed by contracting muscle, though whether NO diminishes or enhances such effects is unclear. Conceivably, during states of great hyperactivity of muscle, hydroxyl free radicals formed from NO and superoxide could lead to muscle damage — in such circumstances, NOS inhibitors might display therapeutic actions. ❑

References

1. Moncada, S., Palmer, R. M. J. & Higgs, E. A. *Pharmac. Rev.* **43**, 109-142 (1991).
2. Dawson, T. M. & Snyder, S. H. *J. Neurosci.* **14**, 5147-5159 (1994).
3. Kobzik, L., Reid,, M. B., Bredt, D.S. & Stamier, J. S. *Nature* **372**, 546-548 (1994).
4. Huang, Z. *et al. Science* **265**, 1883-1885 (1994).
5. Nathan, C. *FASEB J.* **6**, 3051-3064 (1992).
6. Verge, V. M. K. *et al. Proc. Natn. Acad. Sci. U.S.A.* **89**, 11617-11621 (1992).
7. Herdegen, T. *et al. J. Neurosci.* **13**, 4130-4145 (1993).
8. Bredt, D. S. & Snyder, S. H. *Neuron* **13**, 301-313 (1994).
9. Dinerman, J. L. *et al. Proc. Natn. Acad. Sci. U.S.A.* **91**, 4214-4218 (1994).
10. Kobzik, L. *et al. Am. J. Respir. Cell. Molec. Biol.* **9**, 371-377 (1993).
11. Nakane, M., Schmidt, H. H. H. W., Pollock, J. S., Forstermann, U. & Murad, F. *FEBS Lett.* **318**, 175-180 (1993).

****

Questions:

1. Where is nitric oxide formed in the body?

2. How do researchers study nitric oxide?

3. In what new area of the body has nitric oxide been identified?

Answers are at the end of the book.

While searching for clues to the mystery of why so many people have trouble learning to read, scientists discovered a new neurobiological phenomenon: sex differences in phonological processing by the brain. What this means is men and women differ in how their brains match a symbol to a sound. This intriguing information is riding a wave of questions regarding whether men and women actually differ in the way their brains process stimuli and how they both arrive at the same results. Scientists hope to use this knowledge to explain why women are better able to recover from learning disabilities than men. Although the evidence is compelling, its true meaning is still open to interpretation.

S/He-Brains

Men are from Mars, women from Venus. What's more, women do phonological processing with both their right *and* left inferior frontal gyri.

by Sarah Richardson

For most kids, learning to read is just a question of practice. But an estimated 20 percent of Americans have persistent trouble converting letters on a page into sounds. Recently a brain-imaging study pinpointed where that seemingly magical conversion takes place. That was an important result: a first step toward untangling the neurological basis of why so many perfectly smart people have trouble learning to read. But what grabbed headlines was the study's second result: women do it differently from men.

The authors of the study, pediatrician Sally Shaywitz and her husband, neurologist Bennett Shaywitz, weren't looking for sex differences; as codirectors of the Yale University Center for the Study of Learning Attention, they've been studying reading disabilities for a long time. For their most recent study, though, the Shaywitzes used the most sophisticated new brain-imaging technique available—functional magnetic resonance imaging (fMRI). Like other brain-imaging techniques, fMRI detects tiny shifts in blood flow that indicate which parts of the brain are more active. But fMRI, says Sally Shaywitz, is more precise than other techniques—which is how she and her husband ended up discovering sex differences in phonological processing, or letter-to-sound conversion.

The Shaywitzes studied 19 men and 19 women, all neurologically normal. As the subjects lay in the fMRI scanner, they were tested on four language-related tasks—each requiring them to compare two patterns on a video screen, and each more complex than the last. In the first task, the subjects decided whether two series of lines were going in the same direction. In the second, they decided whether two strings of uppercase and lowercase letters matched in both letter and case. In the third, they decided if two nonsense words—say, *slote* and *roat*—rhymed. This task tested the ability to match symbol to sound—the skill that most people with reading disabilities are deficient in. The final task tested understanding: the subjects had to decide, for example, whether the words *senator* and *congressman* belong in the same category.

Men and women showed a difference in brain use only on the rhyming task. In both sexes, trying to recognize rhymes led to increased blood flow in the inferior frontal gyrus of the left hemisphere—Broca's area, as it's commonly called—which has long been linked to language ability. In men that was the only active region, whereas 11 of the 19 women also showed activation in the corresponding region in the right hemisphere. The women's brains weren't working any harder overall; the work was just spread out more. "The accuracy was very comparable," Sally Shaywitz says. "It may be that there are just different routes in the brain to get the same results."

Neither the Shaywitzes nor anyone else, though, knows for sure what the difference they detected means. Does it relate in any way to the observation that women are less likely to lose language ability when they suffer left-hemisphere strokes? To the fact that women tend to do better on tests that measure verbal fluency?

The women in the Shaywitz study, it's worth emphasizing, didn't do "better" than the men, they just did differently—and they did differently only on the rhyming task. On the fourth task, which required them to understand the meaning of words, their brains worked the same as the men's. And even on the rhyming task, only 11 of the women were different from the men; the other 8 were not. So it's not even clear that the difference in brain use detected by the Shaywitzes is a natural difference between the sexes; conceivably it could result from differences in upbringing.

The truth is, we're no closer to answering the timeless questions—why men are less likely than women to ask for directions, more likely to vote Republican, and so on. On the other hand, we may be a little bit closer to understanding reading disabilities. Women tend to have more success at overcoming such disabilities—is it because they process letters into sounds on both sides of the brain? Says Shaywitz: "We're going to investigate that possibility." ❑

****

Questions:

1. What sophisticated brain-imaging technique was used in this study to observe brain activity?
2. Which area of the brain showed differences when tested in either men or women given a rhyming task?
3. What impact will this research have in understanding reading disabilities?

Answers are at the end of the book.

Magnetic resonance imaging (MRI) and positron-emission tomography (PET) are new imagery methods that are revolutionizing the science of cognitive psychology. New theories will have to be developed based on the results of imaging studies as scientists can now visualize the working brain as it views the world and thinks about it. With these imaging techniques, it is possible now to show neuronal activity being localized in areas of the brain during such visual functions as color processing and motion detection. Brain activity can also be measured in people as they are presented with sensory stimuli as a way of understanding how the region-specific activity changes with each cognitive process. Knowledge of the emerging functional anatomy of the brain resulting from the use of PET and MRI studies will open the progress of brain research and allow cognitive psychologists to see the mind more clearly than ever before.

Seeing the Mind

by Michael I. Posner

The microscope and telescope opened vast domains of unexpected scientific discovery. Now that new imaging methods can visualize the brain systems used for normal and pathological thought (1), a similar opportunity may be available for human cognition. Some of the data generated with these imaging techniques fits with current ideas, but much of the new information will require new theories. Here, I review three areas of cognitive psychology in which our ideas are changing as a consequence of these new results: the localization of mental operations (2), the identification of separate brain control systems (attentional networks) (3), and the convergence of sensory input and mental imagery in the same brain areas (4). New theories based on these findings are likely to change our interpretation of the meaning of what has already been seen and to suggest new experiments.

The major new anatomical tools all image aspects of blood vessel function that reflect nearby neuronal activity. The two most successful so far are positron emission tomography (PET) (1) and magnetic resonance imaging (MRI) (5). Although the fundamental links between blood flow and neuronal activity remain obscure, it is clear that the measurement of cerebral blood flow can tell us where neurons are more or less active in comparison to a control condition. Of course, an increase in neuronal activity may signal either inhibitory or excitatory synaptic events, so these measures must also be related to the efficiency of task performance (2). In addition, blood flow lags behind neural activity in time, requiring the use of other methods with time resolution in the millisecond range if changes in anatomy are to be traced dynamically during mental tasks. Methods based on noninvasive electrical and magnetic measurements (6) will be important in studying these dynamic changes when they can be related to the emerging functional anatomy from PET and MRI studies.

It is a popularly held belief in psychology that the cognitive functions of the brain are widely distributed among different brain areas. Even though the organization of the nervous system suggests that sensory and motor functions are localized to specific brain regions, the failure of phrenology and difficulties in locating memory traces for higher functions have led to a strong reaction against the notion of localization of cognitive processes (7). Nevertheless, imaging studies reveal a startling degree of region-specific activity. The PET studies show clearly that such visual functions as processing color, motion, or even the visual form of words occur in particular prestriate areas (4). This localization extends beyond sensory systems. When thought is analyzed in terms of component mental operations (2), a beautiful localization emerges. In word-reading studies, words activate specific posterior visual areas that are not affected by consonant strings, and specific frontal and temporo-parietal

Reprinted with permission from the author, Michael I. Posner and *Science*, October 29, 1993, Vol. 262, pp. 673-674.

areas are active when subjects are required to indicate the use of a noun (for example, hammer-pound) or its classification into a category (hammer-tool) (5). The brain activation accompanying this form of semantic processing is illustrated in the figure, in which the subjects were asked to determine the use of visually presented nouns (8).

The PET method used in the figure required averaging among subjects within a normalized brain space and revealed the surprising fact that the anatomy of this high-level cognitive activity was similar enough among individuals to produce a focal average activation that was both statistically significant and reproducible (2). Functional MRI studies allow examination of brain activity in each individual in relation to their own brain structure and are likely to extend the PET findings so that individual differences in brain anatomy can be related to mental operations (5).

Imaging has also taught us how our brain pays attention to certain stimuli. A common method in functional imaging is to compare a passive control condition with active conditions, such as generating the meaning of a word (9). Some brain areas are activated when subjects have to perform active tasks irrespective of their content. These areas appear related to focal attention, as described in cognitive studies. In the figure, when subjects are first generating the meaning of nouns, there is activity in a midfrontal area called the anterior cingulate gyrus, but after practice with a list, the activity is reduced and returns when a new list is given. Areas in and around the anterior cingulate are active during tasks that involve words, spatial objects, or motor learning and do not seem to be specific to any particular domain of activity (4, 10). During generation of novel word meanings, in addition to the anterior cingulate, two left cortical areas known to be related specifically to word processing are also activated. In general, attentional networks appear to comprise a number of subcortical areas that may differ depending upon whether their function is to orient to sensory stimuli, maintain the alert state, or carry out a variety of executive operation (3). These anatomical results open up the possibility of progress in many topics: what brain structures are involved in awareness, in voluntary control of information storage, in motor responding, and in maintaining current goal states (11).

The new brain imaging techniques have revealed a convergence of what we see and what we think; thinking about a telephone activates some of the same brain areas as seeing a telephone. Specific brain areas are activated when people are presented with sensory stimuli. These activations are in the areas one would expect on the basis of the type of sensory stimulation involved (4). For example, activations have different locations within the primary visual cortex when visual events are presented at different retinal locations, as would be expected from many visual mapping studies (12). When the visual stimuli utilize color or motion, prestriate areas become active that correspond to what would be expected from studies of cellular recording in monkeys (4).

When subjects are instructed to attend to color or motion, there is an increase in activation in the same prestriate areas that process these sensory dimensions (4). Moreover, if subjects are asked to create a visual image based on their remembered knowledge of a visual form, areas in the visual system also show increased activation (13). These findings support the general idea that processes initiated internally from instructions can activate the same sensory areas where these computations are performed on actual sensory events. The finding that imagery and perception share some of the same neural machinery was anticipated by many cognitive theories, but the topic had been subject to seemingly endless dispute before this evidence from brain imaging methods was available (14).

Understanding of how voluntary attention affects the activity within sensory-specific cortex requires an analysis of the brain circuits by which higher level instructions influence sensory areas. Because of the relatively long delay between neural changes and the changes in blood vessels, it is useful to relate the functional anatomy method of PET and MRI to time-dependent measures involving surface or depth recording of electrical or magnetic potentials (6). These methods allow a precise measurement (in milliseconds) of changes between experimental and control conditions and can be used for tracing the circuitry of a particular mental activity. Methods of recording electrical and magnetic activity outside the head have improved in recent years and we now know that there are specific generators of these signals. This knowledge has already helped to spur developments relating the two types of measures (15).

The ability to study the human brain by physiological methods is likely to transform our understanding of what the brain does. If the neural systems used for a given task can change with 15 minutes of practice as in the figure, how can we any longer separate organic structures from their experience in the organism's history? We must be able to trace the changes in the brain that occur with experience. Individual genetic makeup and learning together shape brain structure. We now have methods to understand how this takes place and what it means for the limits of human potential. ❑

References and Notes

1. M. E. Raichle, in *Handbook of Physiology; Section I, The Nervous System, Vol. V. Parts 1 & 2; Higher Functions of the Brain,* F. Plum, Ed. (American Physiological Society, Bethesda, MD, 1987), vol. 5, pp. 643-674.
2. M. I. Posner, S. E. Petersen, P. T. Fox, M. E. Raichle, *Science* **240**, 1627 (1988).
3. M. I. Posner and S. E. Petersen, *Annu. Rev. Neurosci.* **13,** 25 (1990).
4. M. Corbetta, F. M. Meizin, S. Dobmeyer, G. L. Shulman, S. E. Petersen, *J. Neurosci.* **11,** 2383 (1991); S. Zeki *et al., ibid.,* p. 641.
5. G. McCarthy, A. M. Blamire, D. L. Rothman, R. Gruetter, R. G. Shulman, *Proc. Natl. Acad. Sci. U.S.A.* **90,** 4952 (1993); S. E. Petersen, P. T. Fox, A. Z. Snyder, M. E. Raichle, *Science* **249**, 1041 (1990).
6. G. R. Mangun, S. A. Hillyard, S. J. Luck, *Attention and Performance XIV* (MIT Press, Cambridge, MA, 1993), pp. 219-244.
7. For a review of this history, see E. R. Kandel, J. H. Schwartz, T. M. Jessell, *Principles of Neural Science* (Eisevier, New York, ed. 3, 1991), pp. 5-17.
8. M. E. Raichle *et al., Cerebral Cortex,* in press.
9. S. E. Petersen, P. T. Fox, M. I. Posner, M. Mintun, M. E. Raichle, *J. Cog. Neurosci.* **1**, 153 (1989).
10. J. V. Pardo, P. J. Pardo, K. W. Haner, M. E. Raichle, *Proc. Natl. Acad. Sci. U.S.A.* **87**, 256 (1990).
11. M. I. Posner and M. K. Rothbart, in *The Neuropsychology of Consciousness,* A. D. Milner and M. D. Rugg, Eds. (Academic Press, London, 1992), pp. 91-112.
12. P. T. Fox, F. M. Miezin, J. M. Allman, D. C. Van Essen, M. E. Raichle, *J. Neurosci.* **7**, 913 (1987).
13. G. Goldenberg *et al., Neuropsychologia* **27**, 641 (1989); S. W. Kosslyn *et al., J. Cog. Neurosci.* **5**, 263 (1993).
14. S. W. Kosslyn, *Image and Mind* (Harvard Univ. Press, Cambridge, MA, 1980); Z. W. Pylyshyn, *Psychol. Rev.* **88**, 16 (1981).
15. J. P. Wikswo, A. Gevins, S. J. Williamson, *EEG and Clinical Neurophysiol.* (Eisevier Scientific Press, Dublin, 1993), in press; D. M. Tucker, *ibid.*

16. This research was supported by the Office of Naval Research Contract N:0014-89-J3013 and by a grant from the James S. McDonnell Foundation and Pew Memorial Trusts to the Center for Cognitive Neuroscience of Attention.

****

Questions:

1. PET and MRI are acronyms for what new imaging techniques?

2. What do PET and MRI imaging techniques measure in the brain?

3. How will cognitive psychologists use these imaging techniques to study the brain?

Answers are at the end of the book.

Serotonin is considered a neurotransmitter in the brain and has been implicated in a wide range of behavioral disorders involving moods, the sleep cycle, eating and the sex drive. In addition, serotonin-related problems can exist in schizophrenia, depression and Parkinson's disease. Serotonin activity in the brain is clustered in the brainstem with nerve-fiber terminals distributed within the central nervous system. It has been estimated that each neuron exerts an influence over as many as 500,000 target neurons. Researchers in the field have learned that when the serotonin neurons are active, tonic motor activity is enhanced and sensory-information processing is inhibited. This explains why repetitive acts associated with obsessive-compulsive disorders such as hand-washing and pacing further increase serotonin activity in the body. Likewise, physical exercise that involves repetitive motions such as running may actually help to reduce depression and other disorders by increasing the activity of serotonin neurons.

Serotonin, Motor Activity and Depression-Related Disorders

Clues to the origin and treatment of depression and obsessive-compulsive disorders can be found in the role of serotonin neurons in the brain

by Barry L. Jacobs

Prozac, Zoloft and Paxil are drugs that have been widely celebrated for their effectiveness in the treatment of depression and obsessive-compulsive disorders. The popular press has also made much of Prozac's ability to alleviate minor personality disorders such as shyness or lack of popularity. The glamorous success of these drugs has even inspired some writers to propose that we are at the threshold of a new era reminiscent of Aldous Huxley's *Brave New World*, in which one's day-to-day emotions can be fine-tuned by simply taking a pill. Yet for all the public attention that has been focused on the apparent benefits of Prozac-like drugs, the fundamental players in this story — the cells and the chemicals in the brain modified by these drugs — have been largely ignored.

This is partly a consequence of the complexity of the nervous system and the fact that so little is known about *how* the activity of cells in the brain translates into mood or behavior. We do know that Prozac-like drugs work by altering the function of neurons that release the signaling chemical (neurotransmitter) serotonin. Serotonin has been implicated in a broad range of behavioral disorders involving the sleep cycle, eating, the sex drive and mood. Prozac-like drugs prevent a neuron from taking serotonin back into the cell. Hence Prozac and related drugs are collectively known as selective serotonin reuptake inhibitors, or SSRIs. In principle, blocking the reuptake of serotonin should result in a higher level of activity in any part of the nervous system that uses serotonin as a chemical signal between cells. The long-term effects of these drugs on the function of a serotonin-based network of neurons, however, are simply not known.

My colleagues and I have attempted to understand the role of serotonin in animal physiology and behavior by looking at the activity of the serotonin neurons themselves. For more than 10 years at Princeton University, Casimir Fornal and I have been studying the factors that control the activity of serotonin neurons in the brain. I believe these studies provide the linchpin for understanding depression and obsessive-compulsive disorders and their treatment with therapeutic drugs. Our work provides some unique and unexpected perspectives on these illnesses and will serve, we hope, to open new avenues of clinical research.

Serotonin, Drugs and Depression

Communication between neurons is mediated by the release of small packets of chemicals into the tiny gap, the synapse, that separates one neuron from another. The brain uses a surprisingly large number of these chemical neurotrans-

Reprinted with permission from *American Scientist*, September-October 1994, Vol. 82, pp. 456-463

mitters, perhaps as many as 100. However, the preponderance of the work is done by four chemicals that act in a simple and rapid manner: glutamate and aspartate (both of which excite neurons) and gamma-aminobutyric acid (GABA) and glycine (both of which inhibit neurons). Other neurotransmitters, such as serotonin, norepinephrine and dopamine are somewhat different. They can produce excitation *or* inhibition, often act over a longer time scale, and tend to work in concert with one of the four chemical workhorses in the brain. Hence they are also referred to as neuromodulators.

Even though serotonin, norepinephrine and dopamine may be considered to be comparatively minor players in the overall function of the brain, they appear to be major culprits in some of the most common brain disorders: schizophrenia, depression and Parkinson's disease. It is interesting to observe that glutamate, aspartate, GABA and glycine are generally not centrally involved in psychiatric or neurological illnesses. It may be the case that a primary dysfunction of these systems is incompatible with sustaining life.

Serotonin's chemical name is 5-hydroxytryptamine, which derives from the fact that it is synthesized from the amino acid *L*-tryptophan. After a meal, foods are broken down into their constituent amino acids, including tryptophan, and then transported throughout the body by the circulatory system. Once tryptophan is carried into the brain and into certain neurons, it is converted into serotonin by two enzymatic steps.

Serotonin's actions in the synapse are terminated primarily by its being taken back into the neuron that released it. From that point, it is either recycled for reuse as a neurotransmitter or broken down into its metabolic by-products and transported out of the brain. With this basic understanding of serotonin neurotransmission, we can begin to understand the mechanisms of action of antidepressant drugs.

One of the earliest antidepressant drugs, iproniazid, elevates the level of a number of brain chemicals by inhibiting the action of an enzyme, monoamine oxidase, involved in their catabolism. For example, monoamine oxidase inhibitors (MAOIs) block the catabolism of serotonin into its metabolite, 5-hydroxyindole acetic acid (5-HIAA), leading to a buildup of serotonin in the brain. Unfortunately, because monoamine oxidase catabolizes a number of brain chemicals (including norepinephrine and dopamine), there are a number of side-effects associated with these drugs. Some interactive toxicity of MAOIs is also a major drawback for their use in the treatment of depression.

Tricyclic antidepressants (so named because of their three-ringed chemical structure), do not share the interactive toxicity of MAOIs. Tricyclics such as imipramine act to block the reuptake of serotonin from the synapse back into the neuron that released it. In a sense, this floods the synapse with serotonin. These drugs are quite effective in treating depression, but they also induce some unpleasant side-effects, such as constipation, headache and dry mouth. This may be due to the fact that tricyclic antidepressants not only block the reuptake of serotonin, but also exert similar effects on norepinephrine and dopamine.

The obvious benefit of the selective serotonin reuptake inhibitors is that their action is effectively limited to the reuptake of serotonin. This probably accounts for the fewer side-effects experienced by people taking SSRIs. Like other antidepressant drugs, the SSRIs have a therapeutic lag. They typically require 4 to 6 weeks to exert their full effects. Claude DeMontigny and his colleagues at McGill University have suggested that one of the consequences of increasing the levels of serotonin in the brain is a compensatory feedback inhibition that decreases the discharge of brain serotonergic neurons. This results in a "zero sum game" in which there is no net increase in functional serotonin. However, with continuous exposure to serotonin, the receptors mediating this feedback inhibition (the 5-HT_{1A} receptor) become desensitized. It is hypothesized that after several weeks this results in progressively less feedback, increased serotonergic neurotransmission and clinical improvement.

Behavior of Serotonin Neurons

Essentially all of the serotonin-based activity in the brain arises from neurons that are located within cell clusters known as the raphe nuclei. These clusters of serotonin neurons are located in the brainstem, the most primitive part of the brain. It is not surprising, then, that serotonin appears to be involved in some fundamental aspects of physiology and behavior, ranging from the control of body temperature, cardiovascular activity and respiration to involvement in such behaviors as aggression, eating, and sleeping.

The broad range of physiology and behavior associated with serotonin's actions is at least partly attributable to the widespread distribution of serotonin-containing nerve-fiber terminals that arise from the raphe nuclei. Indeed, the branching of the serotonin network comprises the most expansive neurochemical system in the brain. Serotonin neurons project fibers to virtually all parts of the central nervous system, from the various layers of the cerebral cortex down to the tip of the spinal cord.

Since the raphe nuclei contain only a few hundred thousand neurons, and the brains of large mammals (cats, monkeys and people, for example) contain hundreds of billions of neurons, the serotonin neurons constitute less than one-millionth of the total population of neurons in the brain. Despite being so vastly outnumbered, serotonin neurons have immense importance: *Each one* exerts an influence over as many as 500,000 target neurons. In this light the activity of individual serotonin neurons bears a closer look.

The behavior of any neuron is typically measured by its electrical activity. One of the more remarkable electrical phenomena is a cell-wide discharge called an action potential, or spike. Action potentials result from the movement of charged particles (potassium and sodium ions) into and out of the cell through specialized channels in the membrane. Action potentials are an important aspect of a neuron's behavior because the rate and pattern of their occurrence is thought to encode information that is conveyed to other cells. They are central to our story because, in the case of serotonin neurons, each electrical discharge results in the release of a small packet of serotonin from the cell, which in turn alters the activity of target cells bearing serotonin receptors. (At this writing, at least 14 types of serotonin receptors are known, each of which contributes to the diverse effects of serotonin throughout the brain.)

Serotonin neurons have a characteristic discharge pattern that distinguishes them from most other cells in the brain. They are relatively regular, exhibiting a slow and steady generation of spikes. Serotonin neurons retain this rhythmic pattern even if they are removed from the brain and isolated in a dish, suggesting that their clocklike regularity is intrinsic to the individual neurons.

One of the first significant discoveries about the behavior of serotonin neurons in the brain was that the rate of these discharges was dramatically altered during different levels of behavioral arousal. When an animal is quiet but awake, the typical serotonin neuron discharges at about 3 spikes per second. As the animal becomes drowsy and enters a phase known as slow-wave sleep, the number of spikes gradually declines. During rapid-eye movement (REM) sleep, which is associated with dreaming in human beings, the serotonin neurons fall completely silent. In anticipation of waking, however, the neuronal activity returns to its basal level of 3 spikes per second. When an animal is aroused or in an active waking state, the discharge rate may increase to 4 or 5 spikes per second.

Since early experimental studies had linked serotonin to so many behavioral and physiological processes, one of our first priorities was to examine the activity of serotonin neurons in animals exposed to a wide variety of conditions. We chose to perform these studies on cats since their brains and their raphe nuclei have been described in considerable detail. While we recorded the electrical activity of individual serotonin neurons in the brain, we exposed cats to various stressors such as loud noise, physical restraint, a natural enemy (a dog), a heated environment, a fever-inducing agent, drug-induced changes in blood pressure, and insulin-induced changes in plasma glucose levels. These conditions would be stressful to any animal and, not surprisingly, each of these stressors resulted in dramatic changes in the animal's behavior and activated its emergency defenses, including certain parts of its autonomic nervous system. Remarkably, however, none of these conditions significantly changed the activity of the serotonin neurons beyond that seen during a spontaneous active state. The results were perplexing. If these powerful stimuli could not perturb a serotonin neuron, what would?

The variable activity of serotonin neurons during different stages of the sleep-wake-arousal cycle provided a clue. Recall that serotonin neurons are silent during REM sleep. One of the fundamental features of REM sleep is paralysis of the major muscles of the body. This is achieved by inhibiting the neurons that control the tone of the body's antigravity muscles. Might there be a relationship between the paralysis that takes place during REM sleep and the silence of the serotonin neurons?

This question was addressed with a relatively simple experiment. The destruction of a discrete part of a cat's brainstem produces an animal that by all criteria appears to enter REM sleep. However, antigravity muscle tone is present in these animals, and consequently they are capable of movement and even coordinated locomotion. (This condition has also been observed in some people who have experienced traumatic brain injuries.)

When such a cat is awake or in slow-wave sleep, the activity of its serotonin neurons is similar to that of a normal cat. When these animals enter REM sleep, however, instead of falling silent, the activity of the serotonin neurons increases. Cats that display the greatest amount of muscle tone and overt behavior during REM sleep also have the most active serotonin neurons. In some instances, the activity of their serotonin neurons reaches the same level as that seen during the normal waking state.

Another experiment provided further evidence for the role of serotonin neurons. When a drug that mimics the action of the neurotransmitter acetylcholine is injected into the same region of the brainstem as in the previous experiment, a condition somewhat reciprocal to normal REM sleep can be produced. These animals are awake, as demonstrated by their ability to visually track a moving object, but they are otherwise paralyzed. As in the normal animal in REM sleep, the serotonin neutrons in these animals are completely silent. In association with our earlier studies, these results suggest that we are closing in on at least one of the roles played by serotonin in the brain: There is clearly a strong relationship between the activity of serotonin neurons and the body's motor activity.

Interestingly, some serotonin neurons tend to become active just *before* a movement begins. Their activity may also occasionally synchronize with a specific phase of the movement — discharging most, for example, during a particular aspect of the quadrupedal stepping cycle. Moreover, the rate of the spike discharge often increases linearly with increases in the rate or strength of a movement, such as an increase in running speed or the depth of respiration.

One final observation provides a noteworthy clue to the function of serotonin neurons. When an animal is presented with a strong or novel stimulus, such as a sudden loud noise, it often suppresses all ongoing behavior, such as walking or grooming, and turns toward the stimulus. This orienting is essentially a "what is it?" response. In such instances serotonin neurons fall completely silent for several seconds, and then resume their normal activity.

Anatomical evidence supports these observations about the activity of serotonin neurons. For one thing, serotonin neurons preferentially make contacts with neurons that are involved in tonic and gross motor functions, such as those that control the torso and limbs. Reciprocally, serotonin neurons tend not to make connections with neurons that carry out episodic behavior and fine movements, such as those neurons that control the eyes or the fingers.

Our observations of the activity of serotonin neurons during different aspects of an animal's behavior lead us to conclude that the primary function of the brain serotonin system is to prime and facilitate gross motor output in both tonic and repetitive modes. At the same time, the system acts to inhibit sensory-information processing while coordinating autonomic and neuroendocrine functions with the specific demands of the motor activity. When the serotonin system is not active (for example, during an orientation response), the relations are reversed: Motor output is disfacilitated, and sensory-information processing is disinhibited.

Brain Cells and Mental Disorders

Although we are far from understanding the precise neural mechanisms involved in the manifestation of any mental illness, a number of studies have linked serotonin to depression. One of the most notable findings is that the major metabolite of serotonin (5-HIAA) appears to be significantly reduced in the cerebrospinal fluid of suicidally depressed patients. Our own studies suggest that serotonin neurons may be centrally involved in the physiological abnormality that underlies depression-related disorders. Recall that serotonin neurons appear to play crucial roles in facilitating tonic motor actions and inhibiting sensory-information processing. If an animal's serotonin neurons are responding abnormally, such that the rate or pattern of their activity is modified, then one might expect that both motor functions and sensory-information processing would be impaired.

Depression is frequently associated with motor retardation and cognitive impairment. If serotonin neurons are not facilitating tonic motor activity, then it should not be surprising that depressed patients feel listless and often appear to require enormous effort merely to raise themselves out of bed. Inappropriate activity during sensory-information processing might also account for the lapses of memory and the general lack of interest in the environment experienced by depressed patients. It might also be worth noting here that the well-known efficacy of REM-sleep deprivation for treating depression is at least partly dependent on serotonin. Since serotonin neurons are usually silent during REM sleep, depriving an animal of REM sleep maintains a generally higher level of activity in the system. Preliminary research in my laboratory suggests that the deprivation of REM sleep also increases the activity of serotonin neurons when the animal is in the awake state.

The activity of serotonin neurons may also be central to the manifestation of obsessive-compulsive disorders. Since our results show that repetitive motor acts increase serotonin neuronal activity, patients with this disorder may be engaging in repetitive rituals such a hand washing or pacing as a means of self-medication. In other words, they have learned to activate their brain serotonin system in order to derive some benefit or rewarding effect, perhaps the reduction of anxiety. Since the compulsive acts tend to be repeated, often to the point of becoming continuous, such activity may provide an almost limitless supply of serotonin to the brain. (The same may also be true for repetitive obsessional thoughts, but this is obviously difficult to test in animals.) Treating obsessive-compulsive disorders with a selective serotonin reuptake inhibitor ultimately accomplishes the same neurochemical endpoint, thus allowing these people to disengage from time-consuming, socially unacceptable and often physically harmful behavior.

Our studies suggest that regular motor activity may be important in the treatment of affective disorders. For example, if there is a deficiency of serotonin insome forms of depression, then an increase in tonic motor activity or some form of repetitive motor task, such as riding a bicycle or jogging, may help to relieve the depression. Indeed, there are various reports that jogging and other forms of exercise have salutary effects for depressed patients. This does not mean that exercise is a panacea for depressive disorders. Since the long-term effects of exercise on brain serotonin levels are not known, the benfits may prove to be transient. On the other hand, exercise may be an important adjunct to drug treatments, and may permit a reduction in the required drug dosage.

Finally, there is another potentially productive avenue for drug intervention in the clinic. It is well established that the activity of sertotonin neurons is under negative feedback control, in which the released serotonin molecules bind to the so-called autoreceptor on the releasing cell and acts to inhibit the cell's activity. Because of this the administration of a metabolic precursor of serotonin, such as the amino acid *L*-tryptophan, cannot significantly elevate the synaptic levels of brain serotonin since neurons compensate by decreasing their activity through the negative-feedback mechanism. However, if *L*-tryptophan treatments are combined with low doses of an autoreceptor-blocking agent, the concentration of serotonin in the synapse might br increased without the use of serotonin reuptake inhibitors. This might prove helpful to patients who have adverse reactions to these drugs and might also circumvent the therapeutic lag of 4 to 6 weeks typically associated with antidepressants. However, because the feedback mechanism is positively correlated with the rate of serotonin neuronal activity, our results suggest that drugs that block the autoreceptor might be ineffective in quiescent, lethargic or somnolent patients. Conversely, these drugs may be most effective in active patients or those activated by artificial means.

Conclusion

Our research raises the issue of why the manipulation of a system that is primarily a modulator of motor activity has profound effects on mood. Aside from recognizing that the raphe nuclei are connected to regions of the brain that are known to be involved in the emotions (such as the limbic system), it is worth noting that a common organizational plan underlies the distribution of serotonin cell bodies and fiber terminals in essentially all vertebrate brains. This implies that the system has been conserved through evolution, and suggests that there may be some adaptive significance to linking mood and motor activity.

Consider the following possibility. We know that emotions play a role in allowing an animal to withdraw from an ongoing sequence of activities to consider alternative paths. When something bad (perhaps even life-threatening) transpires, it seems reasonable to suppress motor activity and to contemplate the available options. To put it another way: If something negative has happened in one's world, it might be counterproductive, or even dangerous, to explore and engage the environment. The most adaptive response is to withdraw and ruminate. In this light, emotions act at a higher level of complexity in the service of effective motor behavior. When one's mood is bright and expansive, on the other hand, it may be profitable to explore new options. Wide mood swings may allow an exploration of a broader spectrum of perspectives and thus may be related to the well-documented relationship between mood disorders and creativity in artists, writers and composers.

As a final note, the brain serotonin system may be involved in some nonclinical aspects of human behavior. Why do some people endlessly engage in rhythmic leg bouncing? What is rewarding about chewing gum? What underlies the therapeutic or reinforcing effects of breathing exercises, and the twirling or dancing movements employed by various cults and religious groups? The reader can probably think of other behaviors that increase serotonin release in his or her brain.

****

Bibliography

Jacobs, B. L. 1991, Serotonin and behavior: Emphasis on motor control, *Journal of Clinical Psychiatry*, 52:17-23.

Jacobs, B. L., and E. C. Azmitia. Structure and function of the brain serotonin system. *Physiological Reviews* 72:165-229.

Jacobs, B. L. and C. A. Fornal, 1993, 5-HT and motor control: A hypothesis. *Trends in Neuroscience* 16:346-352.

Jacobs, B. L., P. J. Gannon, and E. C. Azmitia, 1984. Atlas of serotonergic cell bodies in the cat brainstem: An immunocytochemical analysis. *Brain Research Bulletin* 13:1-31.

Steinfels, G. F., J. Heym, R. E. Strecker, and B. L. Jacobs, 1983. Raphe unit activity in freely moving cats is altered by manipulations of central but not peripheral motor systems. *Brain Research* 279:77-84.

Trulson, M. E., B. L. Jacobs, and A. R. Morrison, 1981. Raphe unit activity during REM sleep in normal cats and in pontine lesioned cats displaying REM sleep without atonia. *Brain Research* 226:75-91.

Wilkinson, L. O., and B. L. Jacobs, 1988. Lack of response of serotonergic neurons in the dorsal raphe nucleus of freely moving cats to stressful stimuli. *Experimental Neurology* 101:445-457.

Questions:

1. What is serotonin?

2. What is the primary function of serotonin in the brain?

3. How are drugs used to alter the function of neurons that release serotonin?

Answers are at the end of the book.

Part Four:

The Mechanical Forces and Chemistry of the Senses

What makes fingertips sensitive enough to read Braille? There are an estimated 17,000 mechanoreceptors within the palm and fingertips that contribute to the tactile sensation. The mechanism behind that sensory capability is dependent on ion channels that are sensitive to mechanical stimulation. Mechanosensitive channels respond to a tactile stimulus by causing a change in the membrane's ion conductance resulting in the depolarization of the membrane. Researchers study their molecular mechanisms using pharmacological agents that act selectively on the channel and by recording the electrical response using a pressure clamp device similar to the original patch-clamp. Never before, however, have ion channels been so clearly defined in explaining tactile sensory functions as we know them.

Mechanoreceptive Membrane Channels

Mechanically sensitive membrane channels may participate in processes as diverse as volume regulation in cells and sound reception in vertebrates

by Owen P. Hamill and Don W. McBride, Jr.

All living organisms appear to have in common the ability to sense and react to mechanical stimulation. So-called mechanosensitive responses range from simple behaviors in unicellular organisms — including a *Paramecium* that reverses its swimming motion after bumping into a barrier — to the more complex behavior of multicellular organisms — including a nocturnal scorpion that is alerted selectively to ground vibrations less than one-billionth of a meter in amplitude. In people, the estimated 17,000 mechanoreceptors in the palm and fingertips of a hand indicate the fundamental importance of tactile sensation. In fact, the refinement of the hand's tactile-grasping ability allowed tool making and most likely promoted the rapid burst in human evolution.

Beyond an organism's ability to sense external mechanical forces, individual cells must be able to sense internal forces that may affect their size and geometry. For example, even a primitive single-cell organism needs the ability to regulate its cell volume when the ionic concentration, or osmolarity, of surrounding fluids varies. Without such volume regulation, an aqueous environment diluted with heavy rainfall, for example, would force a cell to take up water, swell and possibly rupture. This fundamental regulatory mechanism probably arose early in organisms and was conserved during subsequent evolutionary modifications.

Every regulatory mechanism requires an appropriate detection system. In the case of osmotic gradients, a detection system could be based on sensitivity to membrane tension during cell swelling, which could then be used to modulate ionic movement across the membrane. This form of mechanosensitivity, apart from serving a regulatory function, may also have been modified to sense external mechanical stimuli that deform or stretch a cell membrane. Thus they may underlie sensory functions as we know them.

Although the properties of mechanosensory processes and behaviors have intrigued biologists and physicists for well over a century, direct insight into the molecular mechanisms has only been achieved in the last decade. In particular, high-resolution electrophysiological measurements have revealed specific ionic channels in membranes that are gated — opened or closed — by mechanical stimulation. These "mechanogated" channels have been identified in bacteria, fungi, plants and animals. This article focuses on the molecular mechanisms that form a link between a mechanical stimulus and the resulting electrical signal.

From Prods to Pressure Clamps

Initial studies of mechanosensitivity focused on specialized receptors. Such studies revealed that mechanoreceptors can be either *phasic* or *tonic* in their response to stimulation. Phasic receptors — such

Reprinted with permission from *American Scientist*, January-February 1995, Vol. 83, pp. 30-37 as pacinian corpuscles that sense vibration in vertebrate skin and vertebrate hair cells that, for one thing, relay sound waves from the middle ear to the nervous system — respond best to transient stimuli. They stop responding, or adapt, when exposed to constant stimulation. Adaptation maintains the high dynamic sensitivity of a mechanoreceptor but avoids saturation from constant background signals, such as gravity in the case of hair cells that signal head movement. Tonic receptors — such as Merkel cells in mammalian skin and slow-adapting stretch receptors in crustacean muscle — report constant stimulation, and they show little or no adaptation. These characteristics place constraints on the cellular and molecular mechanisms behind mechano-electrical transduction, in which a mechanical event is transcribed into an electrical, or ionic, event that serves as information for the nervous system.

Classical studies of mechano-electrical transduction required a probe to prod the receptor and an intracellular electrode, a fine wire or glass capillary, to record the electrical response of the entire cell. In pacinian corpuscles, for example, mechanical stimulation leads to a transient increase in a receptor membrane's conductance to sodium and potassium ions, and somewhat to calcium ions. In a resting receptor, one not responding to a stimulus, sodium ions are concentrated in the extracellular fluid, and potassium ions are concentrated in the intracellular fluid. Because a resting nerve-cell membrane has a higher permeability to potassium ions compared with sodium ions, the ionic concentration differences result in a net efflux, or outward flow, of potassium ions from a cell, which generates a positive charge on the outside of a cell and a negative charge on the inside, or a voltage across a membrane. When channels in the membrane open to both sodium and potassium, the result is a net inward current that decreases the voltage across the membrane, or *depolarizes* it.

Opening channels to sodium and potassium ions depolarizes a membrane through a combination of chemical and electrical forces. Potassium ions tend to move out of a cell to their region of lower concentration, but the positive voltage outside a cell repels the positively charged potassium ions. Positively charged sodium ions, on the other hand, are pulled inward because of the lower concentration and by the negative voltage inside a cell. When channels open to both sodium and potassium ions, the combined chemical and electrical forces create an unbalanced ionic flow: The entering sodium ions outnumber the exiting potassium ions. This creates a net inward current, and the positively charged sodium ions offset some of the negative charge inside the cell, thereby depolarizing it.

In principle, a variety of ion-transport processes could mediate the conductance increase to sodium and potassium ions. For instance, a membrane pump could bring in sodium ions and expel potassium ions. Such pumps, however, are relatively slow and produce small currents. Many mechanoreceptors activate rapidly (in less than one-thousandth of a second) and produce nanoamperes (billionths of an ampere) of current. Such fast-acting, large-amplitude currents are more consistent with gated membrane channels, which can be opened or closed on a millisecond time scale and have transport rates as high as 10 million ions per second.

Although gated ion channels were first hypothesized in the 1950s to explain bioelectrical transmission, only the development of the patch-clamp technique in the mid-1970s by Erwin Neher and Bert Sakmann of the Max Planck Institute (for which they were awarded the Nobel Prize in 1991) directly demonstrated individual gated channels in biological membranes. In the patch-clamp method, a glass "patch" pipette is pressed gently against a cell membrane, and suction draws a membrane patch, about two-millionths of a meter across, into the pipette's tip. After a membrane patch seals to the glass pipette, it is isolated electrically from the rest of the cell so that minute currents (less than picoamperes, or trillionths of an ampere) can be recorded in a patch.

Depending on the type of cell and the channel of interest, appropriate stimulation can be applied to a patch to activate specific ion channels in a membrane. For example, voltage-gated channels, those that open or close when exposed to a specific voltage, can be activated by controlling the voltage of a patch of membrane; specific chemicals can be included in the patch-pipette solution to activate transmitter-gated channels, those that are gated by chemicals; or suction or pressure can be applied to a patch to activate mechanogated channels. In the original patch-clamp technique, suction or pressure was applied by mouth or syringe. More recently, we developed a pressure-clamp arrangement that provides the rapid and precise suction or pressure waveforms that are necessary to describe the dynamic properties of mechanogated channels.

Tapping Toad Eggs

Ideally, one would like to identify and characterize single mechanogated channels in sensory mechanoreceptors such as pacinian corpuscles or hair cells. So far, however, these specialized mechanoreceptors have proved inaccessible to patch-clamp studies because they are either embedded deeply in tissue, surrounded by accessory structures, or have specializations, such as cilia, that prevent direct sealing of a pipette on their mechanosensitive membrane. Fortunately, a wide variety of nonsensory cells, which are accessible to patch-clamp recording, also express mechanogated channels.

The first indication of mechanogated channels in nonexcitable, or non-neuronal, cells came from patch-clamp studies of volume regulation in red blood cells of frogs. Hamill showed that osmotic swelling in these cells activates specific potassium- and chloride-ion-selective membrane channels. Soon after that discovery, Falguni Guharay and Frederick Sachs of the State University of New York at Buffalo demonstrated that stretching the membrane of skeletal muscle, via suction on a pipette, activates channels for positively charged ions. Since these initial reports, a variety of stretch- and volume-sensitive ion channels have been found in the membranes of cells from organisms in all the living kingdoms, and these channels exhibit variability in conductance, ion selectivity and gating sensitivity (whether they are turned on or off by mechanical stimulation). These different properties presumably reflect the different physiological functions of the channels.

A particularly convenient preparation for patch-clamp studies of mechanogated channels in nonsensory cells exists in oocytes, female gametes or reproductive cells, of the toad *Xenopus*. This very large single cell, about one millimeter in diameter, possesses an even distribution of mechanogated channels over the cell surface, leading to about 10 million channels. This channel displays a number of properties that are similar to the mechanosensitive conductances in specialized mechanoreceptors, such as pacinian corpuscles. Although the functional role of a mechanogated channel in oocyte physiology has not been determined, it may play a role in sensing membrane tension during volume regulation, cell-growth regulation or embryogenesis or perhaps even in detecting mechanical perturbations of the membrane by sperm during fertilization.

Opening, Closing and Blocking

An ion channel is characterized in large part by two features: open-channel properties — including conductance, selectivity and channel blocking by ions and drugs — and gating properties — including the probability of being open or closed and the conditions that alter that probability. In principle, a channel could be mechanosensitive because either its conductance or its gating is altered by mechanical stimulation.

These two possibilities can be distinguished by a pressure-ramp protocol, the application of pressure that increases smoothly in the shape of a ramp. In a patch containing many channels, a pressure ramp generates an electrical response that saturates, or reaches a maximum value, before the pressure reaches its peak level. This indicates that the mechanosensitive current is not induced mechanically as a nonspecific leak or because of damage to the cell. (In fact, the membrane current returns to normal levels after removing the pressure stimulus.) Furthermore, applying the same protocol to smaller patches, containing fewer channels, reveals single events that underlie mechanosensitive currents. These events exhibit two states, either open or closed, and switching between states provides mechanosensitivity. The conductance, or current, associated with a single event remains unchanged as pressure increases. So saturation of a response in a multi-channel patch arises from a dynamic equilibrium between the opening and closing of a finite number of channels. Like many mechanoreceptors, the toad-oocyte mechanosensitive channel allows both sodium and potassium ions to pass when it is open.

One difficulty in studying mechanogated channels has been the relative lack of specific pharmacological agents that act selectively on them. Discovering such drugs could prove crucial to revealing the function of these channels in different types of cells. Amiloride and its various structural analogues, however, block mechanosensitive channels in oocytes that are open. Interestingly, amiloride also blocks fertilization, volume regulation and cell proliferation in various cell types, as well as mechanotransduction in hair cells.

Vibration and Adaptation

Different mechanisms may confer mechanosensitivity on a channel. These mechanisms may be classified as direct or indirect, depending on how the mechanical energy is coupled to the channel-gating mechanism. Direct-coupling mechanisms arise from intramolecular arrangements of the channel molecule and viscoelastic elements inside a cell; biochemical reactions and diffusion do not limit the speed of such processes, which may develop in less than milliseconds. In indirect-coupling mechanisms, by contrast, the channel itself is not mechanosensitive. Instead, a second messenger — such as calcium ions, cyclic adenosine monophosphate (cAMP) or fatty acids — diffuses from a mechanosensitive enzyme to the channel, thereby invoking a longer latency, from several milliseconds to seconds. Using a pressure clamp, we found that the oocyte mechanosensitive channel can turn on in as few as 100 microseconds, which is comparable to the fast turn-on times in vertebrate hair cells. Such rapid activation behavior in an oocyte and a hair cell is more consistent with direct mechanical-gating mechanisms.

In many cases, the turn-off time of a channel is just as important as the turn-on time. These two times, in combination, determine a mechanotransducer's ability to follow oscillations in mechanical energy, such as vibrations. Although oocyte mechanogated channels turn off a little slower than they turn on, the channel should be able to follow mechanical stimulations up to a frequency of about 200 hertz. It is not clear when an oocyte might experience such high-frequency mechanical oscillations, but one possibility is during fertilization. During that process, sperm make a high-frequency bombardment on an oocyte, and an oocyte undergoes rapid changes after a single sperm enters, thereby preventing polyspermy. Mechanosensitive channels may be involved in this process, but little is known presently about the mechanically induced events associated with fertilization.

Beyond high-frequency vibration, a mechanoreceptor may experience a constant stimulus. Like pacinian corpuscles and hair cells, mechanogated channels in both oocytes and skeletal muscle adapt rapidly to a sustained stimulus. Both suction and pressure produce similar levels of adaptation in oocyte mechanosensitive channels, suggesting that a change in membrane tension, not pressure *per se*, activates the channel. Moreover, the channels do not become inactivated or desensitized by a constant stimulus. A mechanosensitive channel adapted to constant suction can be reactivated by increasing the suction.

A simple model suggests a possible mechanism behind adaptation in a mechanosensitive channel. Imagine that a channel is held closed by an elastic element, such as a spring, connected in series with a viscous element, such as a dashpot, which is basically a shock absorber. When a channel is closed, the spring is coiled rather tightly and the dashpot's piston is bottomed out, or pressed against the bottom of the pot. When a channel opens, the spring stretches, and the dashpot remains bottomed out at first, because a damping element responds more slowly than an elastic element. If the stimulus continues long enough, a channel closes — the sign of adaptation — as the spring recoils and the piston extends in the dashpot. If a second larger stimulus is then applied, the spring stretches and the dashpot remains unmoved. Eventually, the spring would recoil and the dashpot's piston would extend to the end of its range, or top out, which would allow the channel to close, leading again to adaptation. This would explain adaptation, reactivation and a second bout of adaptation. When the stimulus is removed, a channel would recover, letting the spring and dashpot return to their original states.

Adaptation in *Xenopus* oocytes resembles that in hair cells. The major difference lies in their sensitivity to calcium ions: Hair cells require an inward flow of calcium ions for adaptation, whereas mechanosensory adaptation in oocytes is calcium independent. In addition, *Xenopus* mechanosensory channels can switch from an adapting to a nonadapting mode if subjected to repeated, long pulses of suction. The channel current's adaptation decreases with successive pulses, and eventually the channel loses all mechanosensitivity. This behavior probably arises from a disruption of connections between the membrane and the cytoskeleton because of mechanical stresses from patch recording and stimulation. In fact, photomicrographs show that gentle suction on a patch leaves the membrane in close proximity to the underlying cytoplasm, and that maintained suction leads to a distinct separation between the membrane and the dark-pigmented cytoplasm, presumably delineating the cytoskeleton of an oocyte. Although this cellular behavior represents artifacts related specifically to patch-clamp recording, the possibility exists that in different cells the presence or absence of membrane-cytoskeleton interactions may be programmed genetically to provide either adapting or nonadapting responses. Furthermore, such membrane-cytoskeletal interactions may be subject to modulation in a cell either during development or as a result of pathological conditions.

Models of Mechanosensitivity

Direct mechanical gating of a mechanosensory channel has two fundamental requirements. First, the channel needs a mechanism by which tension can be applied specifically to a channel molecule. Second, the channel molecule must be sensitive to the applied tension. Recent evidence indicates that tension may be exerted on a channel by a variety of mechanisms.

In bacteria, for example, a mechanosensitive-channel protein has been purified, cloned and sequenced. When this channel is reconstituted in cytoskeleton-free liposomes, essentially membrane-bound bags, it forms mechanosensitive channels. So tension is most likely exerted on the channel by elastic devices in the membrane itself. The reconstituted channels, though, display relatively low mechanosensitivity, and they do not show adaptation to sustained stimulation, suggesting that interactions with viscoelastic elements in the extracellular or cytoskeletal domains may be required to confer high sensitivity and adaptive behavior on mechanosensitive channels.

The concerted involvement in mechanoelectrical transduction of all three domains — extracellular, membrane and cytoskeletal — is perhaps best exemplified by the specialized vertebrate hair cell. A hair cell's upper surface includes an array of slender hairs called a hair bundle. One "hair" is called the kinocilium, which is built like a true cilium, and the others are called stereocilia, which are built like the microvilli of the intestinal wall. In this system, extracellular "tip links," or fine filaments, interconnect individual stereocilia of the hair bundle. When a bundle is deflected, the tip links deliver, or focus, tension on mechanosensitive channels in the membrane of the stereocilia. Furthermore, hair-cell adaptation supposedly involves an intracellular myosin-actin motor that actively resets the tip-link tension after adaptation, essentially taking up the slack, which poises membrane channels on the threshold of opening. Therefore, this motor is the basis for the exquisite sensitivity and adaptive behavior of a hair cell's mechanosensitive channels.

There is no information on the structure of mechanosensitive channels in oocytes and skeletal muscles or on how they are activated. Nevertheless, we have a speculative scheme for the elements behind mechanoelectrical transduction in these channels. To begin with, we assume that any mechanical stimulus applied to a cell — whether it be an overall stretching or a localized perturbation — will distort the extracellular matrix as well as the membrane-cytoskeleton complex. Molecules that span the membrane may form part of a chain that focuses external mechanical energy on specific membrane-channel molecules. For example, proteins called integrins connect extracellular matrix molecules to cytoskeletal proteins. A variety of cytoskeletal proteins may also be links in this chain. In particular, proteins of the dystrophin-spectrin family are attractive candidates, because they lie directly beneath the membrane and have been shown to interact with a number of membrane glycoproteins. Interestingly, dystrophin has actin-binding domains that may generate active tension in the cytoskeleton, possibly priming mechanosensitive channels through a mechanism similar to the motor process of hair cells. Over the next few years, investigators should identify the exact molecular elements that are involved in this proposed chain of events.❑

Bibliography

Brownell, P., and R. D. Farley. 1979. Detection of vibrations in sand by tarsal sense organs of the nocturnal scorpion, *Paruroctonus mesaensis. Journal of Comparative Physiology* 131:23-30.

French, A. S. 1992. Mechanotransduction. *Annual Review of Physiology* 54:135-152.

Guharay, F., and F. Sachs. 1984. Stretch-activated single ion channel currents in tissue-cultured embryonic skeletal muscle. *Journal of Physiology* 352:685-701.

Hamillp, O. P. 1983. Potassium and chloride channels in red blood cells. In *Single-Channel Recording*, ed. B. Sakmann and E. Neher. New York: Plenum Press.

Hamill, O. P., A. Marty, E. Neher, B. Sakmann and F. J. Sigworth, 1981. Improved patch clamp techniques for high-resolution recording from cells and cell-free membrane patches. *Pflügers Archiv* 391:85-100.

Hamill, O. P. J. W. Lane and D. W. Jr. McBride. 1992. Amiloride: a molecular probe for mechanosensitive channels. *Trends in Pharmacology Sciences* 13:373-376.

Hamill, O. P., and D. W. Jr., McBride. 1994. Molecular mechanisms of mechanoreceptor adaptation. *News in Physiological Sciences* 9:53-59.

Hudspeth, A. J. 1989. How the ear's works work. *Nature* 341:397-404.

Hudspeth, A. J., and P. G. Gillespie. 1994. Pulling springs to turn transduction: adaptation by hair cells. *Neurons* 12:1-9.

McBride, D. W. Jr., and O. P. Hamill. 1992. Pressure-clamp: a method for rapid step perturbation of mechanosensitive channels. *Pflügers Archiv* 421:606-612.

Morris, C. 1990. Mechanosensitive ion channels. *Journals of Membrane Biology* 113:93-107.

Neher, E., and B. Sakmann. 1992. The patch clamp technique. *Scientific American* March: 44-51.

Sukharev, S. I., B. Martinac, V. Y. Arshavsky and C. Kung. 1993. Two types of mechanosensitive channels in the *Escherichia coli* cell envelope: solubilization and functional reconstruction. *Biophysical Journal* 65:177-183.

Wang, N., J. P. Butler and D. E. Ingber. 1993. Mechanotransduction across the cell surface and through the cytoskeleton. *Science* 260:1124-1127.

****

Questions:

1. What are the two types of mechanoreceptors?
2. How do mechanoreceptors respond to touch?
3. How do pharmacological agents help in the study of ion channel function?

Answers are at the end of the book.

Looking for sound in all the wrong places? That's what neurobiologist John Middlebrooks and his colleagues at the University of Florida thought when they went looking for a spatial map in the auditory cortex of the brain responsible for sound perception and found nothing. This meant that the higher auditory system didn't seem to have a map of space with which to pinpoint sound similar to that used by the visual system. What they did learn was that sound location was conveyed not by the spatial arrangement of the neurons but by their temporal pattern of firing. In other words, the brain may be tapping out the location of a sound by changing the pattern of firing like the neuronal equivalent of the Morse code. What responds to the auditory neurons code for location is still unknown, but with the right experiments they should be able to track down the brain's method for computing sound location in the not too distant future.

Neurons Tap Out a Code That May Help Locate Sounds

by Marcia Barinaga

The brain is a history of past experiences, as well as a computer, weighing data and plotting our next move. But it's also an atlas: The cerebral cortex contains a collection of neuronal maps showing where sensations come from. If you cut your left index finger, for example, your brain registers the location of the hurt by firing neurons in the part of the map of skin surfaces corresponding to that finger. If someone throws you a ball, you know where to reach out to catch it, because neurons in the appropriate part of the brain's visual map register the ball's approach.

The brain is also skilled at pinpointing sounds, but so far researchers looking for a spatial map in the auditory cortex—the brain area responsible for sound perception—have come up empty-handed, prompting speculation that the brain may have other tricks for registering the location of sounds. Now, John Middlebrooks and his colleagues at the University of Florida Brain Institute may have confirmed those speculations. They report their finding that neurons in the auditory cortex convey information about a sound's location not by the neurons' spatial arrangement, but by their temporal pattern of firing. In effect, the brain may be tapping out the location of a sound in a neuronal equivalent of Morse code. That possibility is "what makes the paper interesting," says auditory neuroscientist Eric Young of Johns Hopkins University in Baltimore. "The higher auditory system, where sound perception probably happens, doesn't seem to have a map of space...So we are looking for other ideas."

From the beginning of his career, Middlebrooks has been on a quest for auditory maps. In the early 1980s, he worked as a postdoc with Eric Knudsen at Stanford University. Knudsen and Mark Konishi of the California Institute of Technology had already found a map of auditory space in the barn owl. That map, however, wasn't in the auditory cortex but in the superior colliculus, a part of the brain stem that controls the reflex by which the owl turns its head to face a sound. Middlebrooks found a similar map in the superior colliculus of cats; the combined findings led him to conclude that "the brain apparently knows how to make a [sound location] map."

As a reflex center, however, the superior colliculus has little or no role in perception, and the next step was to move to the cerebral cortex, where sound location is consciously perceived. A number of groups, including Middlebrooks', searched the cerebral cortex for telltale signs of an auditory map—and failed. What they were looking for was a pattern of neurons in a grid, each firing electrical impulses in response to sounds from a particular location and remaining silent when the sounds came from elsewhere. Instead, Middlebrooks found that the neurons in an area of auditory cortex called the anterior ectosylvian sulcus (AES) fired off similar numbers of impulses (known as "spikes" for their appearance on an oscilloscope) regardless of the source of the sound.

Musing over his data, Middlebrooks made a discovery. "Even though spike count didn't seem to vary reliably with sound location, I could see changes in the pattern of the spikes," he says. "There was some order there." Depending on where the sound came from, the spikes might begin earlier or later; their frequency also seemed to change.

Middlebrooks wasn't the first to conclude that information may be contained in the timing pattern of spikes. In 1987, Barry Richmond of the National Institute of Mental Health and Lance Optican of the National Eye Institute reported that some neurons in the visual cortex seem to record information that way. That information probably isn't necessary for recording location, since the visual system has its own spatial maps.

But the auditory cortex doesn't, and Middlebrooks postulated that differences in neuronal firing patterns might code for spatial location of sounds. To see if they do, Middlebrooks, working with postdoc Ann Clock, graduate student Li Xu, and psychophysicist David Green, used a computer-simulated neural network designed to recognize patterns. They made electrical recordings from individual AES neurons in the brains of anesthetized cats while they moved a sound source to different spots in a 360-degree arc around the cats' heads. In each trial, they trained the network by feeding it firing patterns from an individual neuron as well as information about which sound location had produced each pattern. Finally, they tested whether the network could tell from patterns alone where a sound came from.

It could: The neural net located the sound sources more than twice as reliably as would be excepted by chance. A cat's brain does much better, of course, but it also has many more neurons to listen to. "No single neuron is going to be a [perfect] pointer...but it may say 'somewhere over in that direction,'" says auditory neuroscientist John Brugge of the University of Wisconsin. "It may take a number of neurons, all saying that, to actually accurately point in that direction."

Although he doesn't think his finding settles once and for all the issue of how sound location is coded, Middlebrooks says it would be logical for the auditory sense to record information chronologically while the visual system does so spatially. Information enters the visual system in map form, as an image projected on the retina; the brain merely preserves the map as it passes the information along. "But in the auditory system, the ears are mapping frequency, [not location], and you have got to somehow take that information...and compute sound location," says Middlebrooks. Once the brain made that computation, there is no a priori reason why it must store the information in the form of a map.

Despite that reasoning, the jury is still out on whether the Middlebrooks group has indeed found the auditory system's substitute for a spatial map. "The temporal pattern of firing of neurons changes as a function of sound location, so these patterns may constitute codes for sound location," concedes Caltech's Konishi. But, he adds, existence of the information is not enough. To settle the question, researchers must establish that the animal uses it. And that will require finding neurons that respond to the patterns, or experimental ways of altering the firing pattern to observe changes in the animal's response to sound. Those experiments are difficult, but without them the research community is left with the existential question of whether, when the auditory neurons tap out their neuronal Morse code, anyone is listening. ❑

****

Questions:

1. What is the auditory cortex?
2. How does the auditory cortex convey information about the sound's location?
3. How do the ears function in receiving sound?

Answers are at the end of the book.

Nested within the human nose are millions of odor receptors of many varities. With such an extensive repertoire of detectors, many signals have to be sent to the olfactory cortex in the brain to convey the correct smell. How does the information get there in an orderly fashion without getting mixed up? Each signal represents a different odor component and it is filed into little neural junctions called glomeruli which play a crucial role in organizing scent perception. These smell files respond to only a part of the smell molecule's structure, so in order to get the entire picture, an odor has to be distinguished by a pattern of glomeruli it activates. Extensive patterns create the calling card for a given order, which is why the brain can recognize nearly 10,000 scents.

The Smell Files

Our noses impose order on an odorous world by dissecting each smell into components and routing each component into a separate neural file.

by Sarah Richardson

Scent, the most primal of all senses, is arguably the most complex. The human nose contains millions of odor receptors of a thousand kinds—far more variety than is required for color vision or taste. How does the brain make sense of this flurry of signals? An essential first step, two groups of neurobiologists have recently found, is a orderly filing system. As they enter the brain from the different receptors, the hundreds of signals, each one representing a distinct odor component, are sorted into little round files called glomeruli—one type of signal per file.

Odor receptors lie in the mucous membranes high in the nose, on hairlike cilia that are the tips of dendrites—the receiving ends of the olfactory neurons. An individual neuron carries many copies of just one kind of receptor. When a smell molecule binds to a receptor, the receptor sends a signal up the dendrite to the cell body. From there it travels down the neuron's sending arm, the axon. The axon passes through a hole in the bone into one of the olfactory bulbs—twin structures, each roughly the size of a small grape, that are lodged above each nostril on the underside of the brain.

The glomeruli (the word is Latin for "little balls") line the olfactory bulbs. Each one is a neural junction, a place where the axons from roughly 2,000 olfactory neurons meet and pass signals to the dendrites of some 30 mitral cells. The mitral cells then refine the signals and relay them to the olfactory cortex, a higher region in the brain.

That much has been known for some time now. And so it has also been clear for some time that glomeruli must play a crucial role in organizing scent perception. But what hasn't been clear, because the neutral connections hadn't been mapped out, is exactly what that role is. Past studies had shown that different odors activate different subsets of glomeruli; and yet a single glomerulus was also thought to respond to a variety of odors. Is it receiving information from neurons bearing different receptors, or does it receive information from only one kind of olfactory receptor?

By using molecular probes that tag particular olfactory receptors, researchers have recently untangled some of the neural wiring in the olfactory bulbs of rats and shown which wire connects to which junction. Both groups—one led by Richard Axel at the Howard Hughes Medical Institute at the Columbia College of Physicians and Surgeons and the other by Linda Buck of Harvard Medical School—found that glomeruli don't get mixed signals. Instead, each gets a very clear one: only neurons bearing one kind of receptor converge on a glomerulus. In fact, it looks as though there are roughly as many glomeruli as there are kinds of olfactory receptors. A glomerulus for a particular receptor seems to be located in the same place in each of the two olfactory bulbs, and in the same place in all rats.

Each receptor and each glomerulus, Axel explains, responds to just one part of a smell molecule's structure rather than the whole molecule. "What you're doing is essentially dissecting the odorous image," he says. "You deconstruct the image such that the individual components of a given odor—even an odor elicited by a single molecule—will react with different receptors simultaneously." Because different molecules may share some structural features, a single glomerulus may be activated by different odors, even though it receives information from only one kind of receptor.

Any given odor, however, is distinguished by the pattern of glomeruli it activates. Based on this molecular fingerprint, the brain can somehow recognize nearly 10,000 scents, despite having a repertoire of only a thousand olfactory receptors. Glomeruli help us do that not only by sorting the signals but also by enhancing sensitivity, says Buck. The convergence of similar neurons in a single glomerulus helps the brain get enough of a sample to recognize the odor, even if it is present in very small quantities.

This method of identifying odor components may also explain why we can recognize odors we haven't smelled in decades. Without stimulation, says Buck, olfactory neurons will die. Since an individual neuron responds to a component that many odors share rather than to one particular odor, it is stimulated often. That may keep our neurons in fighting trim, Buck speculates, and able to recognize a distinctive pattern of a scent long after it was first encountered.

The most tantalizing question of all, however, remains unanswered. How is the map of activated glomeruli decoded in the brain? The precise positioning of glomeruli within the olfactory bulb probably simplifies the brain's job somewhat. "If the position to which an olfactory neuron projects is fixed," says Axel, "then the brain can use anatomic position as an indicator of odor quality." But to find out if that's what is really going on, he and Buck will have to untangle the nerve knots some more. "The next question is to go one step further into the cortex," says Axel, "and ask how the mitral cells project to the cortex. That might give us some indication of how this map is read." ❑

****

Questions:

1. Where are the signals from odor receptors sent?

2. Why don't signals get mixed up in the glomeruli?

3. How is an odor distinguished by the brain?

Answers are at the end of the book.

Nothing is more satisfying to the senses than the tastes of a fine meal. The tastants responsible for stimulating our senses enter at the taste pore, the opening of the taste bud, and bombard the taste-receptor cells within. Neurons make contact with the taste-receptor cells at the synapse where signals are passed via chemical transmitters and are then simply forwarded to the brain. Sounds simple, but our sense of taste, however, demonstrates more biological complexity when viewed at the molecular level. This is especially true in the sweet-taste and bitter-taste transduction pathways. Interaction of these tastants with receptors activates a diverse family of proteins called G proteins, for GTP-binding proteins. Of interest to scientists is the new G protein isolated from taste cells called gustducin which is specifically expressed there. Gustducin may be involved in the transduction of bitter tastes through a mechanism similar to that used in visual transduction. This draws an interesting parallel between the two senses and provides many new questions regarding the molecular mechanisms involving taste.

The Sense of Taste

The internal molecular workings of the taste bud help it distinguish the bitter from the sweet

by Susan McLaughlin and Robert F. Margolskee

To the epicure, nothing represents the height of refinement better than the diverse tastes of a fine meal. To the biologist, nothing could be more basic than the molecular mechanisms underlying taste. The chemical senses — taste and smell — are the most primitive of the sensory modalities. Before warning labels were invented, people had to rely on their sense of taste to distinguish nutritionally useful substances, which taste good, from potentially toxic substances, which taste bad. Today, people still rely on their sense of taste to avoid spoiled foods and fine-tune their diets to the specific nutritional needs of the moment. A body that has just vigorously exercised may crave salty food; one low on sugar will yearn for something sweet.

Surprisingly, the many taste sensations that delight the epicure are produced from only four types of tastes: salt, sour, sweet and bitter. Like primary colors, these tastes can be blended and combined to create the many shades and hues of flavor. In addition, the sense of smell contributes greatly to the totality of flavor perception. Chewing releases volatile chemicals into the nasal passages; the scents of these chemicals combine with tastes to create even more flavors. That is why that sinfully delicious piece of chocolate cake is so much less delectable when your nose is stuffed during a head cold.

At the molecular level, taste is quite similar to our other senses. A gustatory stimulus activates a taste-receptor cell, which in turn conveys the message to a sensory neuron. A nerve impulse then relays the message to the gustatory centers of the brain, where it registers as a taste. Because taste stimuli are restricted to only four classes, it might be assumed that conveying information about taste would be a simple task. If one were to design a taste system, the obvious solution would be to create four distinct receptor cells, one for each taste category. Each type of receptor cell would then uniquely respond to and transmit information about its own class of stimulus (the so-called "labeled-line" model).

Unfortunately, nature and evolution have not chosen this simple solution. Instead, it appears that each taste-receptor cell is capable of responding to more than one class of stimulus. Therefore, the specificity in the system must occur elsewhere. If a single taste-receptor cell can respond to several types of stimuli, then it might achieve specificity by treating each type of stimulus differently. Indeed, the picture emerging from molecular studies shows that a receptor cells uses a separate processing pathway for each class of taste stimulus.

Increasing the complexity of the situation is the fact that stimuli for a single taste modal-

Reprinted with permission from *American Scientist*, November-December 1994, Vol. 82, pp. 538-545.

ity may themselves come from several different types of chemicals. Bitter tastes, for example, can be elicited by compounds as diverse as caffeine, which is chemically classified as a purine; morphine, an alkaloid; or potassium chloride, a simple salt. Not only does each separate taste modality require its own particular coded pathway, but each taste modality may use more than one processing mechanism.

The mechanism of taste perception has been of interest for thousands of years. Over 2,000 years ago, the Roman philosopher and poet Lucretius suggested "that formed of bodies round and smooth are things which touch the senses sweetly, while all those which harsh and bitter do appear, are held together bound with particles more hooked, and for this cause are wont to tear their way into our senses, and on entering in to rend the body." Although this description of taste would not be well received in scientific circles today, it nevertheless contains a fundamental truth: The shape of a molecule that stimulates taste does indeed determine which taste modality it stimulates.

The scientific understanding of the way that taste cells process different stimuli has certainly progressed since Lucretius's time, but many questions remain. The surprising complexity of what would at first glance appear to be a simple process has stimulated the efforts of biochemists, physiologists and psychophysicists for many years.

Taste Buds

Found primarily on the tongue, taste cells (also called taste-receptor cells) are organized into specialized structures called taste buds — gourd-shaped collections of 50-150 taste cells. Taste-receptor cells are long and spindle-shaped with protrusions at their tips called microvilli. The first step in taste recognition takes place at the taste pore, an opening in the taste bud where the microvilli of the receptor cells contact the outside environment. Molecules that are perceived to have taste, called tastants, enter the taste pore and interact with receptor molecules and channels within the microvillar membrane of the taste-receptor cells.

Microscopic examination of taste buds reveals several distinct cell types: light cells, dark cells, intermediate cells and basal cells. The specific function of each cell type is not yet known. Each may represent a different stage of taste-cell maturation. Basal cells mature into dark cells, which then become light cells.

The life span of an individual taste cell is only 10 days to 2 weeks, so cells within the taste bud are continually being replaced. As a consequence, cells at many different stages of development are present within a single taste bud.

Mammalian taste buds are localized in structures called papillae (Latin for "bumps"). The numbers and types of papillae vary depending on the species. Rats and people, for example, have three types of taste papillae: fungiform, foliate and vallate. Each type of papilla is found on a different area of the tongue. Fungiform papillae are relatively simple structures, generally containing only a single taste bud. Several hundred fungiform papillae are scattered randomly near the tip of the rate tongue. Foliate and vallate papillae are more complex structures, each containing many taste buds. Foliate papillae consist of a series of parallel grooves along the sides of the tongue. Taste buds line the sides of the grooves, with their pores facing the cleft. Taste buds also line the clefts of the vallate papillae, which are circular or horseshoe-shaped trenches at the back of the tongue.

Besides detecting taste stimuli, taste-receptor cells must also convey taste information to the brain. The link between the taste-receptor cell and the brain is the neuron. Neurons infiltrate taste buds and make contact with taste cells at the synapse, a specialized region between the receiving end of the neuron and the sending end of the taste cell. Information is passed from the taste-receptor cell to the neuron by means of chemical transmitters, called neuro-transmitters, secreted by the taste cell into the synapse. Neurons detect these transmitters and react to them with a nerve impulse that is transmitted to the brain. The act of receiving sensory information and then translating it into a signal useful to the nervous system is called sensory transduction.

Salt

Electrically charged ions can pass into and out of cells through specialized protein pores, called ion channels, which are embedded within cell membranes. The quintessential salty taste is the positively charged sodium ion (Na^+) found in common table salt, sodium chloride. Prior to 1981, it was thought that taste cells were impermeable to ions; however, in that year, John DeSimone, Gerard Heck and Shirley DeSimone of the Medical College of Virginia showed that sodium ions could be actively transported across epithelial membranes isolated from dog tongue. They also showed that sodium-ion transport across the membrane could be inhibited by the drug amiloride, which blocks sodium-ion transport through specialized sodium channels. These results suggested that taste cells contain sodium-ion channels. Further studies in 1983 by Susan Schiffman, Elaine Lockhead and Frans Maes of Duke University showed that in humans and rats, the salty perception of sodium chloride is inhibited by amiloride. John DeSimone and his coworkers proposed that sodium ions enter taste cells by passive diffusion through amiloride-sensitive sodium channels.

The interior of an unstimulated taste cell is negatively charged with respect to the external environment. When sodium ions enter the taste cell, the net charge on the inside of the cell becomes less negative — that is, the charge difference between the cell's interior and exterior become less polarized. This depolarization causes the taste cell to release neurotransmitters into the synapse between the taste cell and the adjacent neuron. The nerve responds to the neurotransmitters and relays the information to the brain.

There may be other mechanisms for salt-taste transduction: In some species between 30 and 50 percent of the response to sodium chloride is not blocked by amiloride.

Sour

Whereas many salts taste salty, acids taste sour. Sour-tasting acids can be inorganic compounds, such as hydrochloric acid, or organic, such as lactic acid, but they all share a common characteristic: They all release a positively charged hydrogen ion (H^+). The hydrogen ions released by acids produce the sour taste.

Sue Kinnamon and Stephen Roper of Colorado State University showed that weak acids stimulate mudpuppy (the amphibian *Necturus*) taste cells, and that this stimulation could be inhibited by chemicals that block potassium ion (K^+) channels. This suggested that potassium channels were involved in sour-taste transduction. The proposed mechanism for sour-taste in amphibians is that hydrogen ions block potassium channels, preventing the release of positively charged potassium ions from the taste cell, leading to cell depolarization and subsequent neurotransmitter release.

However, in hamsters, sour-taste transduction may operate through a different mechanism. Timothy Gilbertson, Patrick Avenet, Sue Kinnamon and Stephen Roper of Colorado State University showed that hydrogen ions can center the taste cell through sodium-ion channels and directly cause cell depolarization.

Sweet

The molecular mechanisms underlying salt- and sour-taste transduction involve a direct interaction between the stimuli, which are ions, and the ion channels. However, the mechanisms for the detection of sweet and bitter compounds are more complex.

Sweet stimuli are large and chemically more complex than ions, and cannot enter taste cells through ion channels; rather, sweet stimuli contact specialized receptor molecules on the taste cell's surface. The contact between the sweet stimulus and a surface receptor molecule triggers a flurry of chemical signals inside the taste cell that culminates in the release of neurotransmitter.

The internal chemical signals generated in response to external stimuli are called second messengers. Second messengers include ions (frequently the calcium ion), large fatty molecules (for example, the lipids phosphatidyl inositol and diacylglycerol), and small molecules (for example, the cyclic nucleotides cAMP and cGMP). Second-messenger pathways involve a long string of chemical reactions where the product of one reaction is the activator for the next, until finally the last event in the chain stimulates a particular cellular activity or cellular output.

Particularly important regulators of second messengers are the GTP-binding proteins, or G proteins. There are two main types of G proteins: small G proteins (for example, Ras, a G protein implicated in carcinogenesis), and the large heterotrimeric G proteins.

Heterotrimeric G proteins make up a diverse family of proteins involved in signal transduction in many different cell types, including smooth-muscle cells, neurons and sensory cells.

The heterotrimeric G proteins are made up of three subunits; alpha, beta and gamma. The specificity of the G protein is largely dependent on the alpha subunit, which interacts with the cytoplasmic face of a specific receptor molecule embedded within the cell's membrane. The alpha subunit ties the G protein to the next component in the transduction pathway. In effect, the alpha subunit determines in which chemical pathway the G protein lies.

Different G-protein alpha subunits activate different second-messenger pathways in the cell. Stimulatory G proteins (G_S and G_{olf}) stimulate the production of the cyclic nucleotide cAMP by activating the enzyme adenylyl cyclase, which makes cAMP from ATP. Inhibitory G proteins (G_{i1}, G_{i2} and G_{i3}) inhibit adenylyl cyclase and thus block the formation of cyclic nucleotides.

G proteins and cyclic nucleotides play an important role in the first steps in the visual-transduction pathway. The rod and cone cells of the retina contain G proteins called transducins ($G_{t\text{-}rod}$ and $G_{t\text{-}cone}$). Transducins activate the enzyme phosphodiesterase, which converts cGMP into GMP.

G proteins of the G_q class (G_q and G_{14}) activate the enzyme phospholipase C to generate inositol trisphosphate (IP_3) and diacylglycerol (DAG), which serve as second messengers. Other G proteins, such as G_0, interact directly with ion channels.

Although they are involved in many different second-messenger pathways, all G proteins are activated in a similar fashion. When the appropriate molecule binds to the exterior face of a cell-surface receptor, the receptor undergoes a conformational change into its active state; the activated receptor in turn activates the G protein inside the cell. The subunits of the activated G protein then dissociate, and the alpha subunit interacts with the next component in the pathway. Eventually, the alpha subunit reassociates with the other two subunits, and the G-protein complex awaits the next signal from the cell-surface receptor.

Beginning in 1988, evidence began to accumulate that suggested the participation of a G protein in the transduction of sweet tastes. Keiichi Tonosaki and Masaya Funakoshi of Asahi University showed that sucrose caused the depolarization of taste cells. Depolarization was also seen when cGMP or cAMP was injected into the cells, suggesting that the transduction of sucrose involved the formation of cyclic nucleotides. In the same year, Patrick Avenet , F. Hoffmann and Bernd Lindemann of Saarlandes University showed that the injection of cAMP into frog taste cells resulted in cell depolarization owing to a decrease in the flow of potassium ions out of the cell. In addition, they showed that high levels of cAMP in the frog taste cell activated a protein kinase, an enzyme that induces the closure of potassium-ion channels. The enzyme adenylyl cyclase is present at high levels in taste cells, and in 1989, Benjamin Striem, Umberto Pace, Uri Zehavi and Michael Naim of Hebrew University and Doron Lancet of the Weizmann Institute showed that the stimulation of adenylyl cyclase by sweet substances requires guanine nucleotides, suggesting that a G protein (presumably of the G_s family) is involved.

These experiments suggested a possible mechanism for sweet-taste transduction. When a sweet tastant binds to a receptor on the cell surface, a stimulatory G protein (G_s) is activated inside the taste cell. The alpha subunit of G_s dissociates from the G protein complex and stimulates adenylyl cyclase to produce cAMP. The cAMP then activates a protein kinase, which chemically modifies and closes the potassium channel. Closing the potassium channel blocks the exit of positively charged potassium ions, causing the taste cell to depolarize and release neurotransmitter.

Although some of the intracellular molecules that transduce sweet taste have been identified, the nature of the extracellular sweet receptor is currently unknown. Sweet-tasting compounds are chemically very diverse, and include polysaccharides as well as amino acids, proteins and various artificial compounds, such as saccharin and aspartame. It seems unlikely that a single receptor molecule can accommodate all sweet stimuli.

Bitter

Bitter-taste stimuli are the most chemically diverse of all the tastants. Some bitter substances have an affinity for the fatty lipid molecules in the cell membrane — these substances are termed lipophilic ("lipid-loving"). Lipophilic bitter substances may penetrate the membrane directly, whereas hydrophilic ("water-loving") bitter tastants may interact with a cell-surface receptor. In addition, experimental evidence suggests that there may be more than one intracellular signaling pathway involved in bitter transduction.

The detection of bitter substances is extremely important for the survival of an organism; sensitivity to bitter stimuli is a protective mechanism for poison avoidance. Many plant and environmental toxins taste bitter. This provides a strong evolutionary pressure to be highly sensitive to a bitterness. The diversity of mechanisms for detecting bitterness may lead to the ability to detect many different kinds of bitter compounds.

There is strong experimental evidence for one mechanism of bitter-taste transduction. Studies done in 1988 by Myles Akabas, Jane Dodd and Qais Al-Awaqati of Columbia University, and in 1990 by Solomon Snyder and coworkers at the Johns Hopkins University have helped to establish the participation of IP_3 and DAG, and implicate the participation of a G_q protein in bitter taste transduction.

Their experiments suggest that some bitter molecules bind to a cell-surface receptor that activates G_q. Activated G_q activates a phospholipase C to generate IP_3, which is known to stimulate the release of calcium ions from internal cellular stores. Calcium ions can cause cell hyperpolarization by the activation of potassium channels; calcium can also directly activate neurotransmitter release.

Evidence of other bitter transduction pathways is controversial. Studies have linked bitter substances to phosphodiesterase activity. Some experiments, however, suggest that bitter tastants activate phosphodiesterase, while others suggest that bitter substances inhibit phosphodiesterase.

Still a third mechanism has recently been proposed by Yukio Okada, Takenori Miyamoto and Toshihide Sato of Nagasaki University, who have linked bitter substances with chloride-ion secretion, which leads to cell depolarization.

Unique Elements

Many of the molecules involved in taste transduction are common players in the signal-transduction pathways of other cell types. However, taste cells are specialized to detect tastants, so it would not be surprising to find that they use a preferred set of signaling molecules, some of which might be unique to taste cells. In our laboratory, we used molecular biological techniques to identify and clone specific components of taste-transduction pathways. These techniques allowed us to isolate the genes encoding some of the components of taste-transduction pathways.

Our approach has been particularly successful in identifying the G proteins expressed in taste cells. Using a technique called the polymerase chain reaction, we identified three known G proteins expressed more highly in taste cells than in non-taste tongue epithelium G_{13}, G_{14} and G_S). Since these G proteins are expressed in high levels in taste cells, they may be involved in taste transduction.

G_{13} is a member of the inhibitory class of G proteins, which prevent adenylyl cyclase from making cAMP. G_{14} is a G_q-type protein, which leads to the generation of the second messenger IP_3. One of the mechanism for bitter taste transduction implicates a G_q-like protein and IP_3 generation, suggesting that G_{14} might be involved in bitter-taste transduction. The stimulatory G protein G_S stimulates adenylyl cyclase, leading to an increase in cAMP.

In addition to these previously characterized proteins, we isolated a novel taste-specific G-protein alpha subunit. Surprisingly, the inferred amino-acid sequence of this new G protein most closely resembled that of the transducins, the G proteins involved in phototransduction. We therefore named this new G protein alpha subunit gustducin, for gustatory transducin. In subsequent experiments we showed that gustducin is expressed specifically in taste cells.

The close relationship of gustducin to the transducins came as a complete surprise. Based on previous experimental evidence, one might expect that the G proteins involved in taste transduction would be G_S (for sweet taste transduction) or G_q

(for bitter taste transduction). The isolation of a taste-specific G protein that resembled the retina-specific transducins suggested that the transduction of chemical taste stimuli might, at least in some ways, be similar to the transduction of light. This is certainly not an intuitive expectation. What role does gustducin play in taste cells? Its close relationship to the transducins provides us with some clues and leads us to a testable hypothesis.

To understand the role of gustducin in taste transduction, it may be useful to examine how the visual transducins transduce light stimuli. Light enters the rod and cone photoreceptor cells of the retina, and activates light-sensitive receptor (rhodopsin) molecules. Activated rhodopsin then activates transducin. The alpha subunit of transducin stimulates phosphodiesterase, which degrades the second messenger molecule cGMP. Decreased levels of cGMP in the rod and cone cells cause positive-ion channels in the cell's membranes to close.

We can make some predictions about how gustducin might operate by comparing its structure to that of the transducins. The regions of the transducin protein that interact with rhodopsin and phosphodiesterase closely resemble the same regions in the gustducin protein. This suggests that gustducin interacts with a membrane receptor in taste cells that may be similar to the visual receptor rhodopsin. In addition, we would predict that gustducin also interacts with a phosphodiesterase in taste cells.

In which type of taste transduction might gustducin be involved? If gustducin activates a phosphodiesterase, then we would expect the levels of cyclic nucleotides (cAMP or cGMP) inside the taste cell to fall following gustducin activation. Since sweet substances are known to increase the amounts of cAMP in taste cells, it is unlikely that gustducin is involved in the detection of sweet tastes.

A more plausible model implicates gustducin in bitter taste transduction. We propose that bitter substances interact with receptors on the taste-cell surface, which in turn activates gustducin. Activated gustducin would then stimulate a phosphodiesterase to degrade cyclic nucleotides. Specialized channels open and allow the passage of sodium, potassium and calcium ions into the cell when cyclic nucleotide levels are low. The entrance of these positively charged ions depolarizes the taste cell, and neurotransmitter is released.

If bitter taste is transduced by mechanisms involving gustducin and phosphodiesterase, or G_q and phospholipase C, how is sweet taste transduced? Activation of G_S proteins leads to an increase in cAMP, which is observed in taste cells exposed to sweet stimuli. Therefore, G_S is the most likely candidate for mediating sweet-taste transduction, and this is consistent with our finding that G_S expression is elevated in taste cells.

A Taste of the Future

To date, the G protein gustducin is the only taste cell-specific protein that has been identified. Taste-transduction pathways may contain other components that are taste-cell specific (for example, taste-receptor molecules), or they may be made up of proteins that also transduce signals in other cell types.

Certainly the most interesting discovery is the close relationship of the taste-specific gustducin to the retina-specific transducins. It suggests that there is a previously unsuspected level of evolutionary conservation at the molecular level between taste transduction and phototransduction. Perhaps taste-cell receptor molecules resemble rhodopsin. The parallels between taste and vision suggest future experiments that would involve mixing and matching taste and visual transduction components to determine whether parts of these two processes are interchangeable.

Many questions about taste transduction remain to be answered, and molecular biology may provide further insight into taste transduction pathways. In all, studies on taste transduction promise to be very sweet for many years to come. ❑

Bibliography

Akabas, M. H., J. Dodd and Q. Al-Awqati. 1988. A bitter substance induces a rise in intracellular calcium in a subpopulation of rat taste cells. *Science* 242:1047-1050.

Avenet, P., and B. Lindemann. 1989. Perspectives of taste reception. *Journal of Membrane Biology* 112: 1-8.

Avenet, P., F. Hoffmann and B. Lindemann, 1988. Transduction in taste receptor cells requires cAMP-dependent protein kinase. *Nature* 331:351-354.

Behe, P., J. A. DeSimone, P. Avenet and B. Lindemann. 1990. Membrane currents in taste cells of the rat fungiform papilla. *Journal of General Physiology* 96:1061-1084.

Derma, 1947. *Proceedings of the Oklahoma Academy of Science* 27:9.

DeSimone, J. A., G. L. Heck and S. K. DeSimone. 1981. Active ion transport in dog tongue: A possible role in taste. *Science* 21:1039-1041.

DuBois, G. E., D. E. Walters, S. S. Schiffman, Z. S. Warwick, B. J. Booth, S. D. Pecore, K. Gibes, B. T. Carr and L. M. Brands. 1991. Concentration-response relationships of sweeteners: A systematic study. In *Sweeteners: Discovery, Molecular Design, and Chemoreception*, ed. D. E. Walters, F. T. Orthoefer and G. E. DuBois. Washington, D. C.: American Chemical Society, p. 261.

Frank, M., and C. Pfaffmann, 1969. Taste nerve fibers: A random distribution of sensitivities to four tastes. *Science* 164:1183-1185.

Gilbertson, T. A., P. Avenet, S. C. Kinnamon and S. D. Roper. 1992. Proton currents through amiloride-sensitive Na channels in hamster taste cells: role in acid transduction. *Journal of General Physiology* 100:803-824.

Gilbertson, T. A. 1993. The physiology of vertebrate taste reception. *Current Opinion in Neurobiology* 3:532-539.

Heck, G. L., S. Mierson and J. A. DeSimone. 1984. Salt taste transduction occurs through an amiloride-sensitive sodium transport pathway. *Science* 223:403-405.

Hwang, P. M., A . Verma, D. S. Bredt and S. Snyder. 1990. Localization of phosphatidylinositol signaling components in rat taste cells: Role in bitter taste transduction. *Proceedings of the National Academy of Sciences* 87:7395-7399.

Kinnamon, S. C. 1988. Taste transduction: a diversity of mechanisms. *Trends in Neuroscience* 11:491-496.

Kinnamon, S. C., and S. D. Roper. 1988. Membrane properties of isolated mud puppy taste cells. *Journal of General Physiology* 91:351-371.

Kinnamon, S. C., V. E. Dionne and K. G. Beam. 1988. Apical localization of K channels in taste cells provides the basis for sour taste transduction. *Proceedings of the National Academy of Sciences* 85:7023-7027.

Margolskee, R. F. 1993. The biochemistry and molecular biology of taste transduction. *Current Opinion in Neurobiology* 3:526-531.

Margolskee, R. F. 1993. The molecular biology of taste transduction. *BioEssays* 15:645-650.

Pfaffman. 1959. *Handbook of Physiology*, Sec I., Vol. I. Baltimore: The Williams and Wilkins Co., p. 507.

Roper, S. D. 1983. Regenerative impulses in taste cells. *Science* 220:1311-1312.

S. K. McLaughlin, P. J. McKinnon and R. F. Margolskee. 1992. Gustducin is a taste-cell specific G protein closely related to the transducins. *Nature* 357:563-569.

Schiffman, S. S. , E. Lockhead and F. W. Maes. 1983. Amiloride reduces the taste intensity of Na^+ and Li^+ salts and sweeteners. *Proceedings of the National Academy of Sciences* 80:6136-6140.

Striem, B. J., M. Naim and B. Lindemann. 1991. Generation of cyclic AMP in taste buds of the rat circumvallate papillae in response to sucrose. *Cellular Physiology and Biochemistry* 1:46-54.

Striem, B. J., U. Pace, U. Zehavi, M. Naim and D. Lancet. 1989. Sweet tastants stimulate adenylate cyclase coupled to GTP binding protein in rat tongue membranes. *Biochemistry Journal* 260:121-126.

Tinti, J. M., and C. Nofre. 1991. Why does a sweetener taste sweet? A new model. In *Sweeteners: Discovery, Molecular Design, and Chemoreception*, ed. D. E. Walters, F. T. Orthoefer and G. E. DuBois. Washington, D. C.: American Chemical Society, pp. 206-213.

Tonosaki, K., and M. Funakoshi 1988. Cyclic nucleotides may mediate taste transduction. *Nature* 331:354-356.

Questions:

1. What are the four types of tastes?
2. What is the location of the taste-receptor cells?
3. What second messengers are involved in the signal transduction of sweet and bitter compounds?

Answers are at the end of the book.

Very little is known about the molecules and pathways that mediate the sensory signals of taste. Although we higher organisms only exhibit four types of taste modalities, salty, sour, sweet and bitter, a myriad of signaling pathways must create the proper strategy for taste transduction to be sucessful. The tastes of sour and salt are believed to be modulated by the entry of H^+ and Na^+ ions through specialized channels of the membrane while sweet and bitter transduction is coupled to G-protein activity. Interestingly enough, a molecule in the bitter signaling cascade called gustducin, has been found to be very similar to the visual transducin used for coupling the light-receptor molecule rhodopsin. Using this information, scientists hope to probe the area of taste transduction and gain insight into the biology of signalling pathways.

A Taste of Things to Come

by Charles S. Zuker

Taste transduction is one of the most sophisticated forms of chemotransduction[1,2], operating throughout the animal kingdom from simple metazoans to the most complex of vertebrates. Its purpose is to provide a signalling response to non-volatile ligands (olfaction provides a highly specific, extremely sensitive response to volatile ligands). Higher organisms have four basic types of taste modality: salty, sour, sweet and bitter. Each of these is thought to be mediated by distinct signalling pathways that lead to receptor cell depolarization, release of neurotransmitter and synaptic activity[3]. Although much is known about the psychophysics and physiology of taste cell function[4], very little is known about the molecules and pathways that mediate these sensory signalling responses. To complicate matters, some taste-receptor cells respond to more than one taste modality, raising important mechanistic questions about signal crosstalk and information encoding and decoding. Now[5,6] Margolskee and colleagues offer us some surprising insights into these signalling cascades.

Sour and salty tastants are believed to modulate taste cell function by direct entry of H^+ and Na^+ ions through specialized membrane channels localized on the apical surface of the cell. In the case of sour compounds, taste-cell depolarization is thought to result from H^+ blockage of K^+ channels, and salt transduction from the entry of Na^+ through amiloride-sensitive Na^+ channels. None of the molecular components of the sour or salty pathway has been identified.

Sweet and bitter transduction are believed to be G-protein-coupled transduction pathways. However, nothing is known about the membrane receptors activated by bitter and sweet compounds — this is surprising, given the tremendous biotechnological and pharmaceutical applications for bitter antagonists and sweet agonists. The transduction to sweet tastants appears to involve cyclic AMP as a second messenger[8-10]. The current view is that a seven-helix transmembrane sweet receptor activates a G_S type G protein, which in turn activates adenylate cyclase. Increased cAMP leads to the activation of protein kinase A and the phosphorylation and blockage of a family of potassium channels. In this regard, cAMP has been shown to mimic the effect of sweeteners, leading to receptor-cell depolarization via a blockage of a potassium conductance. Interestingly, there is also evidence for an amiloride-sensitive conductance in sweet transduction[11].

Bitter transduction appears to make use of a number of intracellular pathways relying on different signalling strategies. For instance, denatonium, one of the most bitter substances known to humans, leads to release of calcium from intracellular stores in some taste-receptor neurons[12]. This suggests an involvement for phosphoinositide-phospholipase C (PLC) signaling in this pathway. But cyclic nucleotide phosphodiesterases (PDE) have also been implicated in bitter transduction. The PLC pathway probably involves the modulation of Ca^{2+}-activated conductances, whereas the PDE pathway may alter the gating of cyclic-nucleotide-modulated ion channels.

The only molecule to be identified so far in the bitter signalling cascade is gustducin[13], a Gα subunit that is highly specific to taste neurons and very similar to visual transducin. Transducin is the Gα subunit in vertebrate photoreceptor cells and is responsible for coupling the light-receptor molecule rhodopsin to a photo-receptor-cell-specific cGMP phosphodiesterase. Active PDE hydrolyses cGMP into GMP and because the light-regulated channels are opened cGMP, the transient reduction in cGMP causes the closing of these cation-selective channels. Thus, the absorption of a photon by the vertebrate photoreceptors causes a brief hyperpolarization of the cell.

In their papers[5,6], Margolskee and colleagues report two intriguing findings on vertebrate taste receptor cells. The first, by Ruiz-Avila *et al.*[5], demonstrates the presence of visual transducin in rat and bovine taste-receptor neurons. The authors describe studies involving immunohistochemistry and the polymerase chain reaction to show that transducin is expressed in some taste receptor cells; moreover, indirect assays provide hints that transducin functions in bitter taste transduction. These results are highly suggestive and will fuel research in this area, but one must be cautious about extrapolating expression studies into functional ones — particularly important here as gustducin and transducin share about 80 percent amino-acid identity. The most rigorous experiment, namely the generation of transducin-knockout mice, is yet to be done, although gustducin knockouts have recently been produced (R. Margolskee *et al.*, unpublished results). The analysis of these animals as well as transducin/gustducin double mutants should provide important insight into the biology of these pathways. Interestingly, if transducin does indeed function in coupling a bitter receptor to phosphodiesterase, then it follows that the receptor in question is likely to share significant homology with rhodopsin.

In the second paper[6], Kolesnikov and Margolskee describe the presence of a cyclic-nucleotide-suppressible conductance in frog taste-receptor neurons. This is something of a novel mechanism, as cyclic nucleotides have never been shown to suppress or inactivate ion channels directly. This is in contrast to cyclic-nucleotide-gated ion channels in which direct binding by cAMP or cGMP opens the channels. Single-channel studies should provide a detailed description of the biophysical properties of these channels and the role of cyclic nucleotides in their regulation. Kolesnikov and Margolskee also show that the electrophysiological responses of some receptor neurons to the two sweetners saccharin and NC-01 (assuming that both taste sweet to frogs) are modulated by IBMX, an inhibitor of phosphodiesterase, suggesting a role for phosphodiesterase in their transduction pathway.

Given that most sweet transduction responses are believed to be mediated through activation of adenylate cyclase, the results presented in these two studies give us a taste of the complexity and excitement of this field. ❑

****

References

1. Avenet, P. & Lindemann, B. *Persp. Taste Recep.* **112**, 1-8 (1989).
2. Margolskee, R. *Bioessays* **15,** 645-650 (1993).
3. Roper, S. D. *A. Rev. Neurosci.* **12,** 329-353 (1989).
4. Gilbertson, T. *Physiol. Vert. Taste recep.* **3,** 532-539 (1993).
5. Ruiz-Avila, L. *et al. Nature* **376,** 80-85 (1995).
6. Kolesnikov, S. S. & Margolskee, R. F. *Nature* **376,** 85-88 (1995).
7. Kinnamon, S. *Trends Neurosci.* **11,** 491-496 (1988).
8. Avenet, P., Hofmann, F. & Lindemann, B. *Nature* **331,** 351-354 (1988).
9. Striem, B., Pace, U., Zehavi, U., Naim, M. & Lancet, D. *Biochem. J.* **260,** 121-126 (1989).
10. Striem, B., Naim, M. & Lindemann, B. *Cell Physiol. Biochem.* **1,** 46-54 (1991).
11. Schiffman, S., Lockhead, E. & Maes, F. *Proc. Natn. Acad. Sci. U.S.A.* **80,** 6136-6140 (1983).
12. Akabas, M., Dodd, J., & Al-Awqati, Q. *Science* **242,** 1047-1050 (1988).
13. McLaughlin, S., McKinnon, P. & Margolskee, R. *Nature* **357**, 563-569 (1992).

Questions:

1. What are the four basic types of taste modalities?
2. How is a response in the receptor elicited by taste?
3. What is gustducin?

Answers are at the end of the book.

Scientists are working on a retinal implant which could conceivably restore sight to some 1.2 million people world-wide who become blind by the disease retinitis pigmentosa. The blindness results when the eye loses its light receptors, namely the rods and cones of the retina. Technology meets the eye in the form of a microchip designed to mimic the retina. A number of very significant obstacles confront this project, however, such as determining the health of ganglion cells in blind individuals and in preparing a safe implant that will be friendly to its environment and communicate effectively with the brain. Despite the problems, this team of researchers believes that the success of a retinal implant is not as far out of view as once envisioned.

Envisioning an Artificial Retina

by Wade Roush

A speck in your eye usually obscures vision. But now researchers in Boston are trying to reverse the usual order of things. They've placed a speck of microelectronics in rabbit's eyes that might — eventually — have a vision-enhancing effect in humans, restoring sight to some blind people.

The speck is a tiny microchip that sits at the back of the eye, and it's designed to translate visual information into electric pulses that are sent to the brain. The device would be a partial substitute retina, a prosthesis for people who have lost the eye's light receptors — the retina's rod and cone cells — through diseases such as retinitis pigmentosa, which affects 1.2 million people worldwide, and age-related macular degeneration, the leading cause of blindness in the West.

There are researchers who doubt this ambitious goal can be achieved. But in March, scientists on the Project for a Retinal Implant (PRI), a joint effort of the Massachusetts Eye and Ear Infirmary, the Massachusetts Institute of Technology (MIT), and Harvard Medical School, completed experiments that offered considerable encouragement. They finished a series of brief in vivo trials of prototype retinal implant components in rabbits. The test devices delivered tiny electrical current to ganglion cells — the nerve bodies on the inside surface of the retina that feed into the optic nerve — and produced measurable activity in the visual cortex of the animals' brains.

These results, say some scientists who are following the research, reflect glimmers of the project's eventual success. "If they can create an effective interface and there are viable ganglion cells, they should be able to produce a partial retinal replacement," says Terry Hambrecht, a physician and electrical engineer who heads the Neural Prosthesis Program at the National Institute of Neurological Disorders and Stroke.

But those are still big "ifs," for even the PRI researchers acknowledge that the uncertainties at the interface between technology and human tissue are legion. "We're facing a number of very significant obstacles, one after the other," says Joseph Rizzo, a neuro-ophthalmologist at the Massachusetts Eye and Ear Infirmary and lead PRI investigator.

The problems fall into three main categories. First, researchers still have to determine how healthy ganglion cells are in the eyes of blind human beings. Second, the scientists need to develop an implantable chip that won't tear or poison the delicate retina. Finally, they must face the biggest problem of all: getting the chip to package visual information in a form the brain can use. And there are certainly researchers who don't believe these problems can be solved. Peter Schiller, a neurophysiologist at MIT, says he doubts any implant will be capable of mimicking retinal output well enough to produce anything resembling normal vision. In the face of such skepticism, the PRI team has not pursued large-

Reprinted with permission from *Science*, May 5, 1995, Vol. 268, pp. 637-638.

scale funding, hoping to first bolster their case with small pilot projects.

What they hope to end up with is a prosthesis that includes a camera, mounted on eyeglasses, that captures visual images with an electronic analog to film known as a charge-coupled device. The device would digitize the images, and the information would be beamed via laser onto the retinal implant. Exploiting an attached electrode array, the chip would convert the laser pulses into a pattern of electric signals. In theory, these signals would then stimulate the nearest ganglion cells, which transmit the information to the optic nerve and the brain, enabling the wearer to perceive an image.

The vision, however, depends on there being enough healthy ganglion cells remaining to transmit the information to the brain. In this area, there is encouraging news. Last October, at the National Institutes of Health (NIH) Neural Prosthesis Conference, a team at another project, the Intraocular Prosthesis Project (IOP) at Johns Hopkins University Hospital in Baltimore, reported on work with three volunteers who had been blinded by retinitis pigmentosa. The retinas of the volunteers were stimulated in one of four quadrants with hand held electrodes, and the people saw phosphenes — points of light — in the corresponding quadrants of their visual fields. This indicated that ganglion cells were still passing signals — and spatially appropriate signals at that — to the optic nerve. "The ganglion cells are functioning about 70% of their normal numbers in patients with outer-retinal degeneration," says Hopkins ophthalmologist Eugene de Juan.

That brings up the next problem: creating an implant that won't harm the retina. This tissue, says John Wyatt, an MIT electrical engineer and Rizzo's co-principal investigator, has the mechanical properties of "about one layer of sticky wet Kleenex." And that doesn't make things easy, as the implant's electrode array, made of polyimide plastic 50 microns thick, has razor-sharp edges. Before insertion, therefore, the array is coated in silicone and bent to match the retina's own curve, so that it rests on the surface without gouging.

In February and March, the researchers wired their prototype electrode array to a thin electric cable and placed it on the retinas of seven rabbits. A stimulus as low as 10 microamps per electrode — a level equal to that envisioned for a permanent, self-contained implant — generated electrical activity in the visual cortex, measured by scalp electrodes. The implants were removed after a few hours, but later this year the PRI group hopes to show that the prototype can be left inside a rabbit eye safely for days or weeks. The group will then graduate to tests in dogs, and — in 5 to 7 years, if all goes according to plan — to humans. The researchers are also testing alternative materials for building and encapsulating the implant, such as low-density plastics, that would be more resistant to corrosive saltwater in the body and would reduce the weight placed on the retina.

The PRI team is putting off the final problem — the electrical interface with the brain — until the device is ready to be implanted in humans. Wyatt explains that optimizing the frequency, timing, and spatial distribution of the electrodes' stimuli to produce useful images will require feedback from experimental subjects; in short, the researchers will have to ask people what they see.

Wyatt is cautiously optimistic that the groundwork laid by the present experiments will pay off. Although ganglion cells come in specialized varieties that respond to many different aspects of visual stimuli — including color, duration, and direction of movement — he says the experiments at Johns Hopkins make it reasonable to suppose that "if you stimulate the ganglion cells in an X pattern, you'll see an X." De Juan and his colleagues at Hopkins are exploring this further. The scientists are stimulating the retinas of their volunteers to determine how much current must be applied to make human ganglion cells fire and how closely together electrodes should be placed to achieve maximum visual resolution.

Not everyone is sanguine about this approach. Schiller, for instance, thinks it's far too simplistic to expect that placing a relatively crude array on top of the highly differentiated ganglion cells will cause the cells to fire in an interpretable pattern — "especially when one appreciates that the retina itself is a very complex computer." Ronald Burde, an ophthalmologist at Albert Einstein College of Medicine in New York, adds that signal-blocking scar tissue is likely to develop wherever a permanent array touches the ganglion cells. "As a useful human tool a retinal implant is at least a half a century away, if not longer," Burde says.

Because of such doubts, the IOP team has not been able to secure — and the PRI researchers have not even applied for — large-scale government funding from sources such as NIH. As a result, all the retinal implant researchers are working with limited budgets drawn mostly from private backers, such as the Lions Club the Research to Prevent Blindness Foundation, and the Seaver Institute of Los Angeles, which specializes in supporting selected high-risk scientific projects. And while all agree that success is a long shot, the implant developers insist that it's not quite as long as doubters suppose. Says Rizzo, "It's our job to do our work well and to show the skeptics that there's reason to reconsider." ❑

****

Questions:

1. What are the light receptors of the eye?
2. What nerve cells will receive the electrical currents sent by the retinal implant?
3. Where in the brain will the visual stimuli be conveyed by the implant?

Answers are at the end of the book.

Part Five:

Important Messengers of the Renal, Endocrine, Cardiovascular, and Reproductive Systems

Small, and quite often forgotten, the thyroid performs a very important function for the body by regulating our basal metabolism. Through the diet, we consume iodine which is readily taken up by the thyroid gland. After a series of conversions, it enters the bloodstream bound to protein in the form or the hormones triiodothyronine (T3) and thyroxine (T4). Once T3 and T4 enter the cells of the body, they control the production of certain gene products that affect our energy requirements. Both the hypothalamus and pituitary help to control the thyroid's release of these hormones. Although disease related to the thyroid and its output can create chronic conditions for the patient, it can be successfully treated.

Thyroid Diseases

Harvard Women's Health Watch

The thyroid occupies a low position in the collective consciousness, yet this small gland influences virtually every cell in the body. The hormones it secretes into the bloodstream play a vital role in regulating our basal metabolism — the rate at which we convert food and oxygen to energy.

How the thyroid works

In some respects the thyroid is like a small factory. It is the site where iodide — the byproduct of the iodine we absorb in food — enters from the bloodstream, is converted back into its original form, and then is welded to proteins to form the hormones triiodothyronine (T3) and thyroxine (T4). T3 enters the cells directly and activates the genes that direct the production of certain proteins. T4 circulates in the blood until T3 is needed; then it is taken into the cells and converted to T3.

The thyroid's output is determined by the hypothalamus, a regulatory region of the brain. The hypothalamus sends thyrotropin-releasing hormone (TRH) to the pituitary — a peanut-sized gland at the base of the brain — which is entrusted with maintaining thyroid-hormone production at genetically determined levels. When the pituitary senses that the supply of thyroid hormones threatens to drop below those levels, it emits thyroid-stimulating hormone (TSH) to trigger the production and release of T3 and T4.

Because the thyroid hormones influence so many types of cells, it's not surprising that when something goes awry we may experience an array of symptoms that seem unrelated. Changes in weight, bowel habits, heart rate, hair growth, temperature tolerance, and menstrual cycles — when they occur together — can signal a thyroid condition.

Our susceptibility to thyroid disease is largely determined by the interaction of our genetic make up, age, and gender. Women, particularly those with a family history of thyroid disease, are much more likely to have thyroid trouble than men. Fortunately, although most thyroid conditions can't be prevented, they respond well to treatment.

When hormone levels are too low

Hypothyroidism, too low a level of the hormones, is the most common thyroid disease in this country, with more than 50% of cases occurring in families where thyroid disease is present. Although as many as one woman in 10 develops hypothyroidism, its symptoms appear so gradually that many are unaware that they have it.

As thyroid hormone levels fall, the basal metabolic rate drops, resulting in a decreased heart rate, and increased sensitivity to cold, sluggishness, constipation, a slowdown in hair and nail growth, and a weight gain that is limited to 10-15 pounds. (Contrary to widespread misconception, hypothyroidism is not responsible for obesity.) Because blood hormone levels are consistently low, the pituitary sends a steady stream of TSH to the thyroid in an attempt to encourage production. This constant stimulation sometimes causes it to enlarge, creating a goiter.

Hypothyroidism may occur in people who take lithium or certain asthma medications, or who have undergone radiation therapy for certain cancers, such as Hodgkin's disease or throat cancer. Iodine deficiency, once a common cause of hypothyroidism, is no longer a problem in this country; most of us get more than enough iodine from salt, bread, milk, and certain prescription and over-the-counter medications.

Hashimoto's thyroiditis is among the most common causes of hypothyroidism in women over 50. It is an inherited condition in which the body produces antibodies against its own thyroid tissue, eventually damaging the gland so much that it can no longer produce enough T3 and T4.

The most accurate test for hypothyroidism measures levels of TSH in the blood. A high TSH level indicates that the pituitary is signalling the thyroid to produce more hormone in an attempt to compensate for too little T3 and T4.

Regardless of the form of hypothyroidism, the condition can be successfully treated with tablets containing thyroxine, or T4, which can be regulated to maintain adequate levels of circulating hormone. Blood levels of the hormones should be checked periodically and the dosage adjusted when necessary, because needs may vary with age, body weight, and use of certain medications. Most patients need to take the thyroxine for the rest of their lives.

When levels are too high

As might be expected, the symptoms of hyperthyroidism are the opposite of those that characterize hypothyroidism — a modest weight loss despite a normal appetite, bouts of mild diarrhea or frequent bowel movements, heat sensitivity, trembling, increased heart rate, and emotional changes. Women with hyperthyroidism have lighter periods and a higher rate of infertility and miscarriage.

There are several conditions that can cause hyperthyroidism, but about half of all cases are the result of *Graves' disease*, which, like Hashimoto's thyroiditis, is hereditary. It is seven to nine times more common in women than in men. It is also an autoimmune disease, but instead of destroying the thyroid tissue, the antibodies in patients with Graves' disease mimic the effects of TSH, triggering thyroid enlargement.

Like those with other forms of hyperthyroidism, people with Graves' disease often take on a "wide-eyed" look, caused by a retraction of the upper lids. Less commonly, the immune reaction responsible for Graves' disease creates swelling in the muscles of the eyes, causing them to bulge, and produces raised, plaque-like areas on the skin of the legs.

In those with a family history of Graves' disease, the condition is most likely to strike between the ages of 20 and 40, sometimes in response to a hormonal change, such as pregnancy. It may also occur after severe emotional stress.

Evaluation for Graves' disease includes a blood test for low levels of TSH and often an iodine uptake test, in which the patient swallows a solution containing radioactive iodine. The physician then uses a scanning devise to measure the amount of iodine that has been absorbed by the thyroid; an elevated level further confirms that the gland is overactive.

Younger patients with mild Graves' disease can be successfully treated with drugs, such as propylthiouracil (PTU) or methimazole, for a period of 6 to 12 months. These drugs make it impossible for the thyroid to use iodine and thus block hormone production. Because antithyroid drugs have several undesirable side effects, many doctors now treat patients initially with radioactive iodine, which destroys thyroid tissue. The dose used to treat hyperthyroidism is much larger than that used to diagnose it, but it passes out of the body through the urine in 48 hours.

Regardless of the method of treatment, high levels of circulating thyroid hormone may remain in the blood for several weeks, so another drug called a beta-blocker is often used to counteract the thyroid hormone's effects on the heart and vascular system. A majority of those treated for Graves' disease will eventually have low thyroid function and require thyroid pills.

Other forms of hyperthyroidism

A single lump, or nodule, which often develops in the thyroids of people who received low-dose radiation in childhood, can sometimes secrete high levels of hormone. When the nodule's production reaches an excessive amount, symptoms of hyperthyroidism begin.

When there are more than one nodule, the resulting condition is called *toxic multinodular goiter*. It usually occurs in people over age 65, and is responsible for about 25% cases of hyperactive thyroid.

Although beta-blockers and antithyroid drugs can alleviate the symptoms caused by thyroid nodules, the only permanent cure is radioactive iodine to destroy the active thyroid tissue. It also eventually necessitates thyroid supplements.

Radioactive iodine therapy has been so successful that thyroidectomy, or removal of the gland, is now rarely used to treat hyperthyroidism. Surgery is usually reserved for patients with goiters that interfere with breathing or swallowing or are unsightly.

Subacate thyroiditis is a less common form of hyperthyroidism caused by a viral infection. In addition to flu-like symptoms, the patient experience an inflammation of the thyroid, which allows excess hormone to leak out into the bloodstream. Although the effects of this form of hyperthyroidism can also be countered with beta-blockers, the symptoms and the underlying infection eventually clear up with treatment. A painless form of thyroiditis occasionally occurs in women who have recently given birth.

Thyroid cancer

Rarely, a lump in the thyroid will have more serious implications. Hoarseness, difficulty swallowing, and a rapid enlargement of a nodule that is not accompanied by symptoms of hypo- or hyperthyroidism may be signs of thyroid cancer. Fortunately, this form of cancer is rare, occurring in 25 people per million, and usually curable. It is most common in those who have received radiation to the head or neck in childhood or who have a family history of thyroid cancer.

When doctors evaluate a thyroid lump, they perform blood tests for hormone levels and autoantibodies to rule out other conditions. A thyroid scan can determine if the nodules fail to take up radioactive iodine and are therefore "cold" or non-functional, which raises the suspicion of cancer. Ultrasound or needle biopsy can indicate whether a cold nodule is a benign fluid-filled cyst or a potentially malignant one. Treatment for thyroid cancer entails removal of all or part of the thyroid, and if cancer has metastasized, therapy with radioactive iodine to destroy the remaining malignant cells is recommended. ❑

****

Questions:

1. What is the function of the thyroid?

2. What is TSH?

3. What is the most common form of thyroid disease in this country?

Answers are at the end of the book.

Secreted by the pineal gland located in the center of the brain is the hormone melatonin. Referred to as the chemical of night in the human body, melatonin production begins at dusk or darkness only to be switched off at the first glimpse of day. The hormone has many roles one of which is as a regulator of the body's internal clock. Travelers, shift workers and blind people would find this drug particularly useful in resetting their biological clocks. Melatonin is also a compound that induces natural sleep, unlike hypnotic sedatives. Some other uses of melatonin are a defense against biologically damaging free radicals, as an age-retarding chemical and coordinator of the hormones involved in fertility. It might sound like a panacea for whatever ails you, but a drug with such wide-range use is sure to become commercially available very soon.

Drug of Darkness

Can a pineal hormone head off everything from breast cancer to aging?

by Janet Raloff

The camera closes in on a young-looking 40-year-old woman examining her troubled face in the mirror. The voice-over asks: "Do you have trouble falling asleep? Are you concerned about unplanned pregnancies? Does jet lag keep you from getting the most out of your vacation or business travel? Want to postpone aging and the chronic diseases that can accompany it?

"If any of this describes you, ask your doctor about MelaTonics. There's a good chance one of our growing family of products can help you. And remember, each is safe, nontoxic, and based on Mother Nature's own recipe."

Sound far-fetched? Today, perhaps. But a host of medical laboratories around the world are addressing these health problems and more with a new class of drugs they hope to introduce within the next decade — some in as little as 3 years.

Each would rely on a synthetic form of melatonin.

Discovered in 1959, melatonin occurs naturally throughout the living world, from algae to primates. Secreted in larger animals by the pineal gland, located in the center of the brain, it is the chemical embodiment of night. In fact, dusk or darkness triggers melatonin production and sunlight inhibits it.

A decade ago, many of melatonin's functions in humans — let alone its mechanisms of actions — remained shrouded in mystery. Today, scientists variously describe its primary role as the regulator of the body's internal clock, a defense against biologically damaging free radicals, an age-retarding chemical, the trigger for sleep, and a coordinator of the hormones involved in fertility.

So it's not surprising that most researchers express skepticism about one another's findings.

Endocrinologist Michael M. Cohen of Applied Medical Research (AMR) in Fairfax, Va., sums up the situation: "There are differences of opinion between scientific groups on the function of this hormone. And as time progresses, all of us will probably learn we had some things right and some things wrong."

Some pharmacies and health food stores already stock melatonin as a sleep aid. However, the chemical's true soporific value is only just emerging, says Richard J. Wurtman, director of the Massachusetts Institute of Technology's Clinical Research Center.

Wurtman began studying what eventually became recognized as melatonin while in medical school, "about 8 million years ago." By 1963, he and his colleagues had written the first paper designating melatonin a hormone. It took another 19 years before they would offer evidence that melatonin can facilitate sleep.

"The trouble was, we had no idea what dose to use," Wurtman recalls. "So we gave people 240 milligrams [mg]" — which seemed reasonable, based on animal studies. While the hormone indeed made people "verrrrry sleeeeeeepy," he notes, it also saddled them with a miserable hangover the next day. The MIT scientists concluded that melatonin offered little promise as a sleeping pill.

Now, however, Wurtman and his coworkers report in May CLINICAL PHARMACOLOGY AND THERAPEUTICS that an eight-hundredth of that earlier dose induces sleep without side effects. And unlike hypnotic sedatives, they find, melatonin does not alter the normal architecture of sleep, including the time and duration of dream phases characterized by rapid eye movements.

"This compound produces natural sleep," Wurtman concludes.

The data suggest that "very tiny doses" of melatonin can combat insomnia, particularly in the elderly. Over time, the pineal gland accumulates deposits of calcium that diminish its melatonin production. That may be one reason many older people have trouble getting a good night's sleep, he says. The low-dose supplement the MIT team now administers appears to foster sleep by raising concentrations of melatonin in the blood from the 10 picograms per milliliter (pg/ml) typical of daytime to the 100 pg/ml that normally circulates at bedtime.

In work published last year, the group found that melatonin can even induce sleep at midday. This suggests, Wurtman says, that the hormone exerts its soporific effect independent of any effects on circadian rhythms, the daily cycles of varying body chemistry, temperature, and wakefulness.

Alfred J. Lewy, a psychiatrist at the Oregon Health Sciences University in Portland, disagrees. He suggests that any sleepiness probably constitutes a "side effect" of melatonin's effects on circadian rhythms. "The most noncontroversial claim that's generally accepted by everyone in the field is the ability of melatonin to phase shift — or reset — the body clock," he contends.

In 1980, Lewy and his coworkers showed that bright light, such as sunlight, affects human melatonin production. This finding led them to use light to treat seasonal affective disorder, or winter depression. Since then, the team has experimented with melatonin to reset the biological clock of travelers, shift workers, and blind people.

"It's when the [body's internal] clock doesn't expect [this hormone] that it's most effective," Lewy says. That's typically the first thing in the morning, to trick the body into thinking dawn has not yet arrived, or during the afternoon, to advance the internal clock to nightfall. "This effect is quite remarkable," he adds, because the 0.5 milligram doses result in concentrations "no greater than those normally [occurring in] the person at night."

Cohen, in fact, has observed no sleepiness among 1,600 Dutch women taking 75 mg of melatonin daily as part of a trial, begun in 1988, examining its prospects for birth control.

Wurtman, who's familiar with the study, concedes the finding is puzzling. However, he says, the doses are so high that they may overwhelm melatonin receptors, which ordinarily undergo stimulation only a few hours a day.

Why add melatonin to the Pill?

Breast cancer risk has been linked to a woman's cumulative exposure to estrogen, which stimulates breast tissue during each reproductive cycle. Exposure really adds up if a woman doesn't become pregnant periodically, shutting down ovulation for a year or so at a time.

Cohen designed an oral contraceptive that suppresses ovulation and reduces the breasts' exposure to estrogen. Together with coworkers in Amsterdam and Jerusalem, he outlines the reasoning behind these pills in the March BREAST CANCER RESEARCH AND TREATMENT. Unlike conventional birth control pills, his contain no estrogen. They do contain the usual progesterone, coupled in this case with melatonin, which can shut down ovulation in seasonally breeding animals.

Cohen says this "breast-cancer-fighting" birth control pill offers a contraceptive success rate comparable to its estrogen-based competition. U.S. trials of the pill have won Food and Drug Administrative approval and will begin after AMR raises the necessary financing.

Meanwhile, Cohen's team is testing a drug that substitutes 75 mg of melatonin for the progesterone normally given daily to postmenopausal women as part of estrogen-replacement therapy. In these older women, progesterone can foster breast cancer and risk factors for heart disease.

Earlier this year, Cohen won a patent for a male contraceptive that pairs melatonin and testosterone, the primary male sex hormone. The pill is designed to stop sperm production and to reduce the risk of prostate cancer. Clinical trials could begin within 2 years.

It was melatonin's ability to enhance immunity that initially intrigued immunologist Walter Pierpaoli of the Biancalana-Masera Foundation for the Aged in Ancona, Italy. His early studies showed that when administered to old mice, the hormone reversed some of the age-related shrinkage in the animals' thymus, an organ that produces certain infection-fighting white blood cells. The treatment also revitalized the animals' immune system.

Last year, Pierpaoli and Vladimir A. Lesnikov of the Institute of Experimental Medicine in St. Petersburg, Russia, offered in the ANNALS OF THE NEW YORK ACADEMY OF SCIENCES (vol. 719) experimental evidence that the pineal gland plays some part in regulating the rate at which the body ages.

The pair cross-transplanted pineals between the brains of young and old mice. Transplanting pineals from 18-month-old mice into 4-month-old mice reduced the younger animals' average life span by one-third. But the reverse procedure enabled 18-month-old mice to live to the age of 33 months — one-third longer than untreated mice.

Pierpaoli interprets these and related findings to mean that "we start aging in the pineal gland." Indeed, he says, "we should think of the pineal as the aging clock" and melatonin as a means by which it translates its timekeeping pulses into body changes. Although "we can dramatically interfere with aging by interfering with the [calcifying] of the pineal," that's impractical at present, he concedes. He suspects that medicine will focus instead on melatonin supplements to compensate for aging pineals.

Russell J. Reiter of the University of Texas Health Science Center in San Antonio also suspects melatonin plays a key role in aging. However, his data suggest it does so by quashing many of the free radicals that, in excess, have been linked to the chronic diseases that tend to accompany aging.

Some cells generate free radicals — molecular fragments possessing an unpaired electron — to destroy unwanted material such as bacteria or aging cells. Drugs, radiation, and other toxic environmental agents also produce free radicals.

Two years ago, Reiter and his colleagues suggested that melatonin might be the most effective antioxidant, or scavenger of free radicals. Since then, he says, "we've tested it in every conceivable system that we can assemble, and melatonin continues to perform as well as or better than any other antioxidant."

In the January MUTATION RESEARCH, the team showed that lymphocytes (white blood cells important to immunity) incubated with melatonin sustained 70 percent less damage from ionizing radiation than untreated cells did. That same month, the group reported in LIFE SCIENCES that rats injected with melatonin before and after exposure to paraquat, a toxic herbicide, showed none of the devastating lung and liver damage seen in unprotected animals.

The signal to produce melatonin goes from the eyes to the biological clock, or suprachiasmatic nucleus (SCN), at the base of the brain. The messenger for that signal is glutamate, Reiter notes, a free-radical-inducing neurotransmitter. Young animals have ample melatonin to neutralize these free radicals, Reiter believes. But "over a lifetime, the glutamate wins, destroying most of the cells in the SCN that generate the melatonin rhythm."

As a result, he proposes, melatonin production starts to drop, leaving the SCN more vulnerable to glutamate attack, which further blunts the output of melatonin. If this explanation for melatonin's age-related decline is validated, Reiter says, melatonin supplements might allow older people to mimic a 20-year-old's natural rhythm — thereby delaying many of the age-related changes fostered by free radical damage.

In a more immediate application, people about to undergo diagnostic X-rays may take a dose of melatonin to protect their healthy tissue.

The one thing on which all melatonin researchers agree is that the hormone is nontoxic, even at fairly high dosages. Toxicology studies by AMR, for instance, indicate that humans could consume 100 times the dosage now used in the Dutch trails without ill effect. Indeed, Cohen argues, on a per weight basis, ordinary table salt is more toxic.

Should consumers hurry to stock up on the supplement? On that, opinions vary widely.

Cohen, whose company expects to become a major

melatonin supplier, says go ahead. Pierpaoli agrees. Even Reiter says, "I take a lot of melatonin."

Others argue that consumers should wait until FDA regulates melatonin. To date, FDA lacks the data it needs to evaluate health claims arising from the new animal and clinical trials. Indeed, the agency cautions consumers that if they take melatonin, "they do so without any assurance that it is safe or that it will have any beneficial effect."

At a minimum, Lewy notes, taking melatonin at the wrong time of day might upset an individual's biological clock. That could prove catastrophic for, say, people driving or physicians conducting surgery. Moreover, Wurtman points out, production of the hormone by unregulated companies increases the risk of impurities tainting the product.

****

Concludes Dan Oren of the National Institute of Mental Health in Rockville, Md., who has studied melatonin, "It's something that I suspect will become available as a commercially regulated drug within a few years. But it is a drug. And its effects have not been adequately studied for people to be responsibly using it on an over-the-counter basis yet." ❑

Questions:

1. What is melatonin?
2. How does melatonin work as a contraceptive in the human body?
3. How does melatonin work as an antioxidant?

Answers are at the end of the book.

Nothing sounds sweeter to couples trying to bear children than the two words, "You're pregnant." But for many couples with infertility the prospects seem dim as month after month passes by with no baby. For this very reason, couples turn to assisted reproduction techniques to help them conceive. Before entering a program, your gynecologist will gather as much information about you and your reproductive health as possible, as well as from your partner. When standard therapy fails, couples can elect for more high-tech approaches to infertility such as in vitro fertilization, gamete intrafallopian transfer and zygote intrafallopian transfer which all involve some method of embryo transfer. Assisted reproduction can be exhausting and expensive but for many its a price worth paying to conceive a baby.

Assisted Reproduction

Harvard Women's Health Watch

Most of us grew up with the notion that it's extremely easy to get pregnant. Needless to say, learning that we may not by able to bear children can come not only as a great surprise but as a major blow.

A few decades ago, the only thing one could do for infertility was to suffer in silence. Today, not only are a host of psychological and social support services available, but several effective treatments as well. Although such high-tech approaches to infertility as in vitro fertilization (IVF), gamete intrafallopian transfer (GIFT), and zygote intrafallopian transfer (ZIFT) may be exhausting and expensive, increasing numbers of couples who are able to keep trying will eventually conceive.

When to consider getting help

Your chance of successful infertility treatment declines rapidly after the mid-30s. If you're in that age range and have been trying to get pregnant for 6 months to a year, you should probably talk to your doctor. (Women under 35 should try for a least a year before seeking help.) Most fertility programs have maternal age limits of 42 or 43, and very few accept women over 45 because their chances of becoming pregnant are so poor.

Your doctor will want to take your medical history, do a physical examination, and draw blood to be tested for hormone levels. Of course, you're not alone in this enterprise, so your male partner will need an evaluation as well.

If your periods are irregular, your blood may be tested at the beginning of your menstrual cycle for follicle-stimulating hormone (FSH), which causes the egg-containing follicle to ripen. High FSH levels indicate that the pituitary gland is producing large amounts of FSH in an effort to get the ovary to release an egg. A blood test for progesterone, which is produced by the empty ovarian follicle (corpus luteum), can be performed later in your cycle to determine if an egg has developed and possibly been released. Another indication that ovulation may be occurring is a mid-monthly decline in body temperature followed by a sustained rise, so you may be asked to record your temperature every morning upon awakening and to plot it on a chart. Not all women experience the temperature drop at ovulation, so you may also be asked to use a home urine test for levels of luteinizing hormone (LH), which rise just before the egg is released, usually between days 10 and 18 of the menstrual cycle.

If these tests indicate that you're not ovulating regularly, your gynecologist may prescribe one of several drugs to stimulate the ovaries, such as clomiphene citrate (Clomid) in pill form, or either human menopausal gonadotropin (Pergonal) or FSH (Metrodin), given as daily injections in the buttocks. All can have unpleasant side effects.

Your male partner will need to donate a semen sample for analysis. If the results are abnormal, he'll be required to repeat the donation, since sample quality can vary widely.

If both your and your partner's fertility evaluations are within normal boundaries, and you have been having intercourse regularly during your ovulatory periods, your doctor will probably suggest a postcoital test. This will require having a vaginal specimen taken within a few hours of intercourse. The sample will be examined under a microscope to determine the number and condition of the sperm that remain and will give your doctor an indication of whether your vaginal mucus is inhospitable to the sperm or you are producing antibodies that disable them.

If your doctor suspects that you are ovulating regularly, but your fallopian tubes are blocked, he or she will probably advise a *hysterosalpingogram* (HSG). In this procedure, dye is infused through the cervix into the uterus and fallopian tubes so that they can be viewed on an x-ray. If the results of the above tests are normal, or if your doctor suspects endometriosis or pelvic adhesions, your physician will probably perform a *laparoscopy* in which a tiny camera is inserted into the abdominal cavity under general anesthesia. This technique allows the doctor to see inside the abdominal cavity and, if necessary, to remove any lesions. (See *HWHW*, September 1994.)

For 10-15% of couples, no reason for infertility can't be found. If less invasive procedures like ovarian stimulation and intrauterine (artificial) insemination aren't effective, these couples, like those for whom the reasons for infertility are known, may benefit from IVF, GIFT, or ZIFT.

Assisted reproduction techniques

The first step in any of the IVF procedures is to "hyperstimulate" the growth of ovarian follicles with Clomid, Pergonal, or Metrodin. You may also be given a gonadotropin releasing hormone (GnRH) agonist, such as Lupron, for several days at the beginning or end of the cycle to suppress your natural hormonal surges. Hyperstimulation causes several eggs to ripen, increasing the number of opportunities for fertilization.

The development of your follicles will be monitored carefully with frequent blood tests and ultrasound. When they ripen, you'll receive an injection of human chorionic gonadotropin (hCG), to trigger the release of eggs.

Retrieving the eggs

About 34-36 hours after you receive the hCG injection, you'll need to return to the clinic so that the eggs can be retrieved. If you're not using donor sperm, your male partner must go along to provide a semen specimen.

The eggs are usually retrieved by a procedure in which the surgeon inserts an ultrasound probe into your vagina. After administering an anesthetic, he or she will pass a hollow needle through the vaginal wall, puncture the ripened ovarian follicles, aspirate the egg-containing fluid, and withdraw the needle. Alternatively, a laparoscopic procedure, which requires general anesthesia, is used for GIFT or ZIFT of for further evaluation of the pelvis.

What happens next depends on the method of embryo transfer you're using. All have similar pregnancy rates.

- *IVF* is usually used for women whose infertility results from obstructed fallopian tubes, endometrosis, or cervical mucus that is inhospitable to sperm. It is also used for some couples with unexplained infertility. The harvested eggs are examined by an embryologist, who places them in a culture dish along with the sperm. When it's evident fertilization has occurred, three or four embryos are transferred via the vagina into the uterus. Using several embryos increases the chance of pregnancy, but it also increases the likelihood of multiple births. Any additional embryos can be frozen for use in a subsequent attempt if the couple wishes.
- *GIFT*. The eggs are retrieved laparoscopically, and then, during the same operation, mixed with the sperm and placed in one or both fallopian tubes. GIFT requires at least one functioning tube and is most often used in cases of unexplained infertility, cervical damage, or mild endometriosis. It is also appropriate for couples in whom the sperm cannot penetrate the cervical mucus.

Unlike IVF, GIFT requires general anesthesia and doesn't allow fertilization to be documented before the egg and sperm are placed in the tube.

- *ZIFT*, an infrequently used assisted reproduction method, is similar to GIFT, except the fertilization occurs in a culture dish before the two-celled embryos, or zygotes, are returned to the fallopian tubes. It is usually reserved for women with severe cervical damage who have at least one working fallopian tube. Both ZIFT and GIFT carry an increased risk of tubal pregnancy.

• *ICSI,* or intracytoplasmic sperm injection, is used for couples in whom sperm count is very low or when motility is extremely poor. In this procedure, a single sperm is injected into an egg, which is then incubated to allow fertilization to take place. The procedure is still experimental, and few clinics are experienced in performing it.

Embarking on assisted reproduction is a major undertaking for any couple. It's wise to determine the cost of treatment and associated expenses, such as travel, in advance. Because insurance coverage is highly variable, check with your insurer to determine how much of the bill you'll have to pay.

****

While you are undergoing evaluation and treatment, it is likely to become the central focus of your life. Like most couples with infertility, you may benefit from seeing a counselor or joining a support group to help you identify options you may not have considered, such as adoption or deciding not to have a child at all. ❑

Questions:

1. What is FSH?

2. In what percentage of couples can no reason for infertility be found?

3. What are the success rates for the various embryo transfer methods?

Answers are at the end of the book.

There is some good news for women who take estrogen replacement therapy to avoid the undesirable effects of menopause. New studies are reporting that estrogen bolsters short-term memory and improves cognitive skills. What is the connection between estrogen and memory? Researchers found in studies with rats that estrogen helps supply an enzyme necessary for the synthesis of acetylcholine, a neurotransmitter crucial for memory. Some of the studies with women taking estrogen show significant improvement in simple memory scores compared with nonusers but others are less encouraging. It may be too early to tell if the results are real or just a placebo effect, but it does tip the scales in favor of estrogen therapy which many women believe has far too many risks already.

Forever Smart

Does estrogen enhance memory?

by Kathleen Fackelmann

Touted in the 1950s as an elixir that would keep postmenopausal women forever feminine and vilified in the 1970s as a cause of cancer, estrogen replacement therapy has had a checkered history.

Yet doctors now know that this drug provides many benefits to women who have gone through menopause. Estrogen shields women from the potentially crippling bone loss of osteoporosis and can slash their risk of heart disease.

Despite the huzzahs given this therapy of late, however, many older women don't take estrogen, perhaps because of a lingering fear of cancer. There's little doubt that estrogen, when taken by itself, magnifies the risk of endometrial cancer. Yet doctors can mitigate this cancer threat by giving estrogen in combination with another hormone called progesterone.

That strategy has done little to assuage another cancer worry, though. Researchers know that a woman's cumulative exposure to her own body's estrogen increases the risk of breast cancer. Unfortunately, studies on the link between estrogen pills and breast cancer have produced conflicting results.

Several research teams, however, now add some more good news to the complex risk-benefit picture of estrogen replacement therapy. Their data suggest it boosts short-term memory and the ability to learn new tasks. In addition, one team suggests that estrogen replacement therapy may counter depression.

"Estrogen has been shown to influence several areas in the brain that are involved in cognition and behavior," says researcher Uriel Halbreich of the State University of New York at Buffalo.

At the same time, a California team reports data hinting that estrogen replacement therapy shields — at least partially — against Alzheimer's disease, a degenerative disorder of the central nervous system.

Is menopause a disease or simply a natural consequence of aging?

Proponents of the disease model point out that at the turn of the century, most women didn't live long past menopause: In fact, only 5 percent of women lived beyond their 50th birthday. However, most U.S. women alive today can expect to spend one-third of their life in that estrogen-deficient state.

Menopause is a dramatic event. Ovarian production of estrogen declines from as much as 300 micrograms per day to almost nothing. Yet postmenopausal women still manufacture up to 20 micrograms of estrogen per day in the liver and fatty tissue.

Women prescribed estrogen replacement therapy have a choice of natural or synthetic estrogen. The most popular product, Premarin, is estrogen derived from the urine of pregnant horses.

With antibiotics, improved diets, and the everyday miracles of modern medicine, humans have expanded their life span. "In a way, we've created this phenomenon of old age," says psychologist Barbara B. Sherwin of McGill University in Montreal.

The theory behind estrogen replacement therapy is that women should be able to avoid the numerous undesirable consequences of menopause, including osteoporosis and hot flashes. Sherwin and other researchers believe it makes sense to supply in pill form the estrogen the body stops manufacturing.

As one ages, it takes longer to remember names and other facts once retrieved without hesitation. While memories of long ago remain sharp, an older person may have difficulty recalling a new telephone number.

Sherwin knew that some research on animals suggested a link between estrogen and memory. In rats, for example, estrogen increases the amount of an enzyme necessary for synthesis of acetylcholine, a neurotransmitter that plays a key role in memory. She wondered whether older women could improve their memory by taking hormone therapy.

Diane L. Kampen, also at McGill, and Sherwin began to study this question by recruiting healthy women age 55 and over who had completed menopause at least 2 years earlier. The duo ended up with 28 women in estrogen replacement therapy and 43 who were not.

To test verbal memory, the researchers read a short prose passage to each recruit. Immediately, and then after a 30-minute delay, the volunteers had to remember as much as they could about the material.

Estrogen users recalled more of the text than nonusers on both tests, the Canadians report in the June 1994 OBSTETRICS & GYNECOLOGY. "In our hands, estrogen enhances verbal memory and helps maintain the ability to learn new material," Sherwin says.

How big a difference is it? Sherwin says it's not a huge effect, but it's enough that estrogen users' performance in the laboratory remains higher that their peers'. In real life, Sherwin says, the difference might come down to this: A postmenopausal woman who starts taking estrogen might recall telephone numbers, instructions, or directions with greater ease.

So estrogen won't turn the postmenopausal woman into a human calculator. On the other hand, who couldn't use a little memory boost now and then? How many times have you gone to a party, been introduced to someone, and then failed to remember the person's name a minute later?

A recent study suggests that hormone replacement treatments may prove an antidote to such forgetfulness. In this study, Jerome Yesavage of the Stanford University School of Medicine and his colleagues decided to study the relationship between estrogen replacement therapy and the ability to recall a name or a specific word.

The team began its study with 72 women between the ages of 55 and 93 who were taking estrogen and 72 women in the same age group who were not. None of the women appeared senile.

To test the women's ability to recall a name, the researchers showed each recruit black-and-white slides of six male and six female faces. The team matched each photo at random with a common first and last name. The women studied each face and the corresponding name for 1 minute. The researchers later shuffled the photographs and projected the faces in random order. The recruits had to write down the name they associated with each face.

Estrogen users performed significantly better than their peers. The researchers discovered that women taking estrogen got 36 percent of the names right, while their counterparts recalled only 26 percent correctly.

The team also administered another simple memory test. It gave recruits a list of 16 common words and asked them to memorize as many as possible in 4 minutes. Later, the women had to write down, in order, the test words they could remember.

On this task, estrogen users did slightly better than nonusers. However, the improvement did not reach statistical significance. The researchers report their finding in the September 1994 JOURNAL OF THE AMERICAN GERIATRICS SOCIETY.

"The implication of this finding, if real, is tremendous — estrogen replacement therapy may confer some protection against memory loss," says epidemiologist Trudy L. Bush of the University of Maryland School of Medicine in Baltimore, who wrote an editorial in the same issue of the journal.

Bush, however, contends that this study, although carefully conducted, is limited. First, she points out, the boost in the ability to remember names, is small. Of dozen names, women on estrogen therapy could remember an average of four. The nonusers recalled about three.

She wonders if some factor other than estrogen use might explain the slight increase in performance enjoyed by those on estrogen therapy. Bush says estrogen users may be healthier than women who do not take the treatment and that better health may explain the improvement in their test scores.

"Right now, the data are pretty weak," Bush says. "I think we need an awful lot more work before we can say anything with any degree of certainty."

Can estrogen reverse some of the ravages of aging? A study by researchers in New York shows that estrogen replacement therapy improves reaction time, verbal ability, mental alertness, and other thinking abilities in postmenopausal women.

Halbreich and his colleagues recruited 34 postmenopausal women who were not taking estrogen and 24 women of childbearing age. The team gave both groups a battery of tests measuring a range of cognitive abilities. The researchers also measures performance on complex cognitive tests. For example, the women took a simulated driving test in which they had to make quick decisions about a variety of road hazards.

In general, the postmenopausal women performed less well then their younger counterparts, Halbreich says. The older women's scores were not abnormal, they simply showed the older women weren't as quick to respond as the younger ones, Halbreich adds.

The researchers then gave the older women estrogen replacement therapy for 60 days. When the team administered the tests again, they found the older women did significantly better on certain tests, including the simulated driving exercise. Indeed, after estrogen therapy, some of the older women posted test scores indistinguishable from the younger women's, Halbreich says.

The researchers also discovered that performance correlated with the amount of estrogen: The higher the concentration of estrogen in the blood, the better the test scores. Halbreich presented the group's results this July at the International College of Neuropsychopharmacology meeting in Washington, D.C.

The findings suggest "a beneficial role of estrogen replacement therapy on some aspects of cognitive functioning," he concludes.

Halbreich cautions, however, that the boost in test scores could be a placebo effect — an improvement not related to the therapy itself — because the older women knew they were getting estrogen. So he and his colleagues plan another study, this one to include a control group of postmenopausal women who will not receive treatment.

The Buffalo-based team also has data hinting that estrogen may help improve mood in older women. Previous research had indicated that postmenopausal women with low concentrations of estrogen may be more vulnerable to depression, perhaps because of a deficit in a neurotransmitter called serotin. Halbreich's data suggest that estrogen modulates serotonin in older women and thus may help regulate mood.

Although estrogen replacement therapy might help ward off depression in older women, it cannot, by itself, banish this disorder, the researchers caution.

The history of estrogen replacement therapy is littered with promises of turning back the clock. Little wonder, then, that researchers remain cautious about these new findings regarding estrogen and thinking ability.

"I think it's hopeful research," Bush says. "But I think we should proceed cautiously." ❑

****

Questions:

1. Where is estrogen produced in the female body?

2. What characterizes menopause?

3. Why do women take estrogen therapy during menopause?

Answers are at the end of the book.

Maleness has always been synonymous with the hormone testosterone. Rightly so, for when the gonads receive a signal to develop testes, they begin to produce the male hormone. But testosterone is not enough. According to researchers, a second hormone must act in concert with testosterone to produce a normal, fertile male. To make a male, the hormone mullerian-inhibiting substance (MIS) is produced in the gonads and released to actively suppress the development of female reproductive tracts. Both male and female possess the MIS gene for producing the hormone. When researchers mutated the MIS gene, it produced some very interesting results. In the end, to achieve the proper sex development requires a balance between these hormones.

Not by Testosterone Alone

There is another hormone that is crucial to manhood—and also to male mousehood. Without it, a male mouse has plumbing problems.

by Sarah Richardson

Making a man is the complex task of culture. Making a male—nature's job—is also none too simple. The first requirement is the sex-determining gene on the Y chromosome, called the SRY gene: it enables a fetus to build testes, which manufacture the sine qua non of maleness, testosterone. But testosterone alone is not enough; a second hormone is required to produced a normal, fertile male. Molecular geneticist Richard Behringer and his colleagues at the University of Texas M.D. Anderson Cancer Center in Houston have shown just how important this hormone is in male mice. When it is absent, the males develop as infertile pseudohermaphrodites—meaning they have the internal reproductive ducts of a female as well as those of a male. "They look like a regular male," says Behringer. "You wouldn't know unless you look inside that something is funny."

In mammals, Behringer explains, male and female embryos are identical in the early stages of development. Each fetus has one set of gonads—reproductive glands that develop into either testes or ovaries. If the gonads receive the signal from the SRY gene, they develop into testes. Other wise they become ovaries.

But this unisex blueprint includes *two* sets of reproductive ducts that extend from the gonads to the embryonic urinary tract. One set of ducts can develop into three female parts: the oviducts (or fallopian tubes in humans), which convey the egg to the uterus; the uterus itself; and the upper portion of the vagina. Very close by lies a second set of ducts that can develop into a male transport system: the epididymis, where sperm cells mature; the vas deferens, the tube that carries mature sperm cells from testes to penis; and the seminal vesicles, which secrete a fluid into the vas deferens that helps rush the sperm on their way. "It's not like the gonads, where you have one pair that have to become one or the other," says Behringer. "You've got two sets of ducts, and you've got to make one set differentiate and one regress, depending on whether it's a male or a female. That takes a bit more coordination."

Making a female is still relatively straightforward, though: in the absence of testosterone the embryonic male ducts wither away. But researchers have long suspected that to make a male, a second hormone must actively suppress the development of the female reproductive ducts, known as müllerian ducts. The hormone, which is made in the gonads, is called müllerian-inhibiting substance, or MIS. Curiously, the MIS gene is on an ordinary chromosome, not on the Y, so both males and females have it.

To pinpoint the role of the MIS gene in embryonic development, Behringer and his colleagues knocked it out: they mutated the gene in embryonic mouse cells, transferred the embryos into a host mother, and bred the resulting offspring until they had mice in which both copies of the MIS gene were mutated. Then they studied the mutants. The mutant females, the researchers found, didn't miss MIS; the gene is turned off in a female embryo anyway, so the mutation had no effect. The mutant males also appeared to be normally equipped. But they had some extras—oviducts, uterus, and the upper bit of a vagina. The lack of MIS had permitted the female ducts to flourish.

Surprisingly enough, that didn't interfere much with the males' conduct. They could still produce normal sperm and copulate with females. What they couldn't do was fertilize a female. Although the mutant males manufactured normal amounts of sperm, 90 percent couldn't father offspring. The presence of the female reproductive duct, Behringer suspects, somehow blocks the path of the sperm cells before they reach their destination. "The two ducts come together," he says, "and you have a plumbing problem."

The same problem has been observed in humans, although it is extremely rare: a man may have fallopian tubes and a uterus, and be sterile, as a result of a mutation in the MIS gene. "What fascinates me is how this delicate balance in normal development is maintained," says Behringer. "At the molecular or embryological level, there's not much difference between men and women. We all start off similar, and a little hormone here or there and you end up being a male or a female." ❑

****

Questions:

1. What gene signals the gonads to develop testes and make testosterone?

2. What function does the mullerian-inhibitory substance have in the male embryo?

3. On which chromosome is the MIS gene found?

Answers are at the end of the book.

Touted as the "fountain of youth," a mysterious hormone has captured center stage in the scientific community for its youth-enhancing qualities. Not much is known about the hormone dehydroepiandrosterone (DHEA) except that it is produced in the adrenal glands and gradually increases from puberty to 25-30 years of age. After that the plentiful hormone declines with advancing age. Volunteers who took DHEA reported an improvement in their general well-being. Researchers are cautious as more information is needed to determine whether DHEA will actually increase a person's life span and help to turn back the biological clock.

Hormone Mimics Fabled Fountain of Youth

by L. Seachrist

When Ponce de Leon set out to find the fountain of youth in the 16th century, his search took him to the swamps of what we now know as Florida.

Today, his scientific counterparts explore the intricacies of the human body in search of life-preserving medicines. As yet, they've had little success. But preliminary results from a study of a plentiful, though poorly understood, hormone could provide the first clues to the whereabouts of that elusive fountain.

A team of scientists from the University of California, San Diego, treated 16 middle-aged to elderly people for a year with either the hormone dehydroepiandrosterone (DHEA) or a placebo. Although the researchers pleaded with attendees at a conference on DHEA and aging in Washington, D.C., this week not to call the hormone "the fountain of youth—it's not a miracle," their results are tantalizing.

The eight study participants receiving DHEA saw a 75 percent increase in their overall well-being, compared to the other volunteers. "It is a small study," says project leader Samuel S.C. Yen. "But there was a marked increase in psychological well-being and ability to cope among these patients."

DHEA, a steroid hormone produced by the adrenal glands, appears in increasing amounts at puberty, reaching its peak between the ages of 25 and 30. Because the hormone slowly declines with age, some scientists refer to it as a biological clock of aging (SN: 6/14/86, p.375).

Animal studies of DHEA suggest that it helps protect against viral infections, heart disease, cancer, and AIDS (SN: 11/2/91, p. 277)

The researchers gave DHEA recipients enough hormone to equal the amount found in the average 30-year-old. The participants filled out questionnaires to record their physical and psychological well-being.

At the end of the study, volunteers taking DHEA reported improved ability to cope with stress, greater mobility, improved quality of sleep, and less joint pain. Men, but not women, experienced an increase in lean muscle mass and a decrease in fat. Neither group perceived any change in libido, and no one reported side effects.

The DHEA group also had higher concentrations of insulin growth factor, a compound that spurs the immune system and normally dwindles with age.

Teasing out the potential risks and benefits of taking DHEA would require "an extensive study in a large population," Yen cautions, adding that there is no evidence to indicate that DHEA increases longevity. He and Etienne-Emile Baulieu of INSERM in Paris have initiated such a project. ❑

****

Questions:

1. Where is DHEA produced in the body?
2. Why is DHEA referred to as an internal biological clock?
3. How much hormone was given to volunteers who participated in the DHEA study?

Answers are at the end of the book.

No pain, no gain. Behind the motto of many athletes who train extensively is the fear that without proper conditioning, they could suffer the consequences of electrolyte, renal and volume disturbances from the most trivial level to the most life-threatening. The metabolic stress of exercise can manifest itself in many forms after the loss of water and electrolytes takes place from the body with complication resulting in muscle cramps, heat exhaustion, and heatstroke. Most serious at the level of the kidney is the risk of developing acute renal failure related to exercise in an underconditioned athlete. Along with better conditioning, athletes must rely on adequate hydration to protect themselves from metabolic distress during extreme exercise and ultimately enhance athletic performance all around.

Exercise-Induced Renal and Electrolyte Changes Minimizing the Risks

by Steven Fishbane, M.D.

In brief: The metabolic stress of exercise can impair the kidneys' ability to maintain volume and electrolyte homeostasis. Minor abnormalities in renal function and plasma electrolyte composition, such as hypovolemia, electrolyte loss, hyperkalemia, and lactic acidosis, may result. In addition, direct and hormonal effects of exercise on the kidneys can lead to proteinuria or hematuria. Rarely, the effects of exercise on the kidneys can be life-threatening, as with heatstroke or rhabdomyolysis. Optimal care not only includes expedient diagnosis and treatment, but also educating the patient about adequate hydration and other preventive measures.

A 34-year-old woman was thrilled to overcome hot, humid conditions to finish a half-marathon. It was a remarkable achievement, because she had trained for only 2 weeks. As she crossed the finish line she was surprised to find that she was not overly tired. Aside from some calluses on her feet, she experienced little discomfort, although she later recalled some lightheadedness when she walked to the sponsor's tables to get a soda after the race.

That night, however, she noticed that the color of her urine was dark red, and by the afternoon of the next day she had stopped urinating. Several hours later she experienced severe pain in both thighs, and after several episodes of nausea and vomiting, she went to a local hospital. She was found to be in acute renal failure, with biochemical evidence of rhabdomyolysis. In spite of her severe renal condition, she eventually recovered uneventfully, never required dialysis, and was discharged from the hospital with near normal renal function.

This case illustrates a rare complication of extreme exercise in an athlete who was not well conditioned. Minor homeostatic disruption, such as moderate hypovolemia, hyperkalemia, lactic acidosis, and hematuria, are much more commonly associated with exercise. Sports medicine physicians need to stay alert for both mild and severe renal and electrolyte changes.

Hypovolemia and Electrolyte Depletion

Deficits in plasma volume and electrolyte stores occur routinely with physical activity. Exercise causes hypovolemia in proportion to the quantity of water and sodium lost in sweat. When a person exercises strenuously on a hot day, the volume of sweat lost can exceed 2 L/hr.[1] In addition, plasma osmolality and sodium concentration are increased because sweat is hypotonic relative to plasma. Both the hypovolemia and hyperosmolality are potent stimuli of thirst, which normally leads to correction of both abnormalities when hypotonic fluids are consumed. During exercise, however, fluid consumption often does not keep up with rapid volume losses.

One circumstance in which severe hypovolemia may develop is in the "weekend warrior"

Steven Fishbane, M.D., "Exercise-Induced Renal and Electrolyte Changes" *The Physician and Sportsmedicine*, Vol. 23, No. 8, copyright 1995, Reprinted with permission of McGraw-Hill, Inc.

who competes in athletics while consuming large quantities of alcohol. The alcohol blocks the effect of antidiuretic hormone in the kidney, preventing maximal concentration of the urine and exacerbating exercise-induced hypovolemia.

The loss of volume during exercise is accompanied by a significant increase in blood flow to muscle and skin.[1] The combination of volume lost as sweat and the redistribution of blood flow to peripheral pools significantly decreases effective circulating volume, stroke volume, and cardiac output.[2] The ability to tolerate this cardiovascular stress depends on the individual's cardiovascular function and level of conditioning. Physical training and heat acclimation both improve the response to hypovolemia.

Hypotension during exercise is prevented by intense vasoconstriction of the splanchnic vascular bed.[3] Alpha-adrenergic receptor blockers (commonly used to treat hypertension) block this protective mechanism and can predispose a patient to hypotensive episodes during exercise. These medications should be used with care in patients who exercise strenuously.

The obligatory loss of volume during exercise is accompanied by losses of electrolytes such as potassium. Potassium depletion can develop after exercise-associated hyperkalemia has reversed with rest. The usual potassium concentration in sweat is 8 mEq/L; therefore, to develop postexercise hypokalemia, total volume losses must be substantial. It has been suggested that potassium depletion may be important clinically, playing a role in the development of heatstroke.[4] Approximately 50% of patients with heatstroke are found to be hypokalemic.[5]

Complications. Loss of water and electrolytes leads to the complications of exercise that are most likely to bring a person to medical attention, including muscle cramps, heat exhaustion, and heatstroke. Exercise-induced *muscle cramps* are common and occur even in well-trained athletes. They can be treated with rest, fluids, and electrolyte replacement. Prevention requires adequate fluid and electrolyte intake before and during exercise.

With *heat exhaustion,* patients experience weakness, headache, nausea, confusion, and faintness. They may have mild hypotension and tachycardia with normal body temperature, but body temperature is often elevated. As with muscle cramps, rest, fluids, and electrolyte replacement are essential. Replacement can be accomplished orally, but intravenous fluids are sometimes required.

Heatstroke is the most serious of the heat-induced syndromes. It occurs following exercise most commonly in older people who have underlying diseases and rarely in young, healthy individuals. This syndrome is a medical emergency and requires immediate recognition and treatment. Its symptoms may be similar to those of heat exhaustion; however, body temperature is elevated, and the patient may present near death. Acute renal failure, rhabdomyolysis, and lactic acidosis may all occur with heatstroke. Treatment involves measures to induce immediate cooling, including infusion of chilled intravenous fluids, massage of the extremities to alleviate vasoconstriction, and use of a fan to promote convective loss of heat.

Prevention. The optimal fluid replacement for volume and electrolyte losses occurring with exercise is controversial. Perhaps the most important principle is to replace as much of the fluid and electrolytes lost as possible, preferably during the exercise itself. Replacing volume losses and preventing hyperosmolality can increase cardiac output, enhance athletic performance, and lower rectal temperature.[6]

At least 1,250 mL of 8% carbohydrate-containing fluids can be absorbed from the intestines per hour.[7] Theoretically it should, therefore, be possible to replace a significant proportion of lost volume in all but the most extreme conditions. In practice, however, the amount of fluid consumed is frequently inadequate. This is particularly true with runners, who usually do not consume more than 500 to 1,000 mL per hour. In part, this may be explained by abdominal discomfort brought on by fluid consumption,[8] however, an important component in many is simply a reluctance to drink fluids during exercise.

Replacement fluids should be hypotonic relative to plasma, to match volume losses as sweat. This will help prevent hyperosmolality. Water by itself, however, may rarely be problematic: Cases of postexercise hyponatremia due to excessive water intake have been reported.

Several sports drinks are marketed as an improved method to replace water and electrolyte losses. The primary ingredients in these drinks are usually water and sugar in the form of corn syrup. The sodium content of typical brands is 10 to 20 mEq/L. Potassium tends to be present in very small concentrations (2 to 4 mEq/L), as are other electrolytes such as calcium and magnesium. The beverages also often contain water-soluble vitamins.

The advantage of these liquids over water is that they provide rapid electrolyte and energy replacement (typically 200 to 300 kcal/L). Carbohydrates in sports drinks may have a benefit because relatively low concentrations (6% to 8%) can serve as both an energy source and a thirst stimulant.[8] There is, however, no clinical proof that these products are necessary. Eating a small snack of almost any type, in addition to drinking water, provides adequate replacement of lost volume and solutes.

Hyperkalemia

Paradoxically, as water and electrolytes (including potassium) are lost from the body during exercise, minor increases in plasma potassium concentration are common. With relaxed walking, increases of approximately 0.3 mEq/L are typical. With more vigorous aerobic exercise, the elevation could reach 0.7 to 1.1 mEq/L.[8-10] With severe exertion (such as marathon running), plasma potassium could increase by as much as 2 mEq/L.[11]

In general, mild hyperkalemia is well tolerated and has no clinical significance. The usual symptoms related to severe hyperkalemia — muscle weakness and cardiac arrhythmias — are only very rarely seen after exercise. One could speculate, however, whether the increased potassium concentrations might play a role in reported episodes of sudden death during exericse.[12] It is possible that in a diseased heart with a predisposition to arrhythmias, hyperkalemia might act as an irritant, thereby decreasing the threshold for malignant ventricular rhythms.

The increased plasma potassium concentration that occurs during exercise results from the release of potassium from muscle cells, followed by delayed cellular reuptake.[13] The slowed reuptake appears to be due to reduced activity of the membrane Na^+-K^+-ATPase. The increased local tissue concentration of potassium may have a beneficial effect, because it leads to vasodilation, which increases blood flow.[14] This provides improved oxygen and nutrient delivery to starved muscle cells and increased removal of accumulated lactic acid.

The increase in plasma potassium concentration generally resolves within 5 minutes after resting.[15] Potassium released from muscle is rapidly reclaimed by cells, so that by the time the physician sees a patient with an exercise-induced injury or illness, the plasma potassium concentration will usually not be elevated.

One situation in which exercise could induce severe hyperkalemia is in patients taking beta-blockers. $Beta_2$ receptor stimulation normally facilitates the movement of potassium into cells. Blockade of these receptors prevents reuptake of potassium released from muscle cells during exercise. Plasma potassium concentrations can increase by as much as 1.5 to 4.0 mEq/L under these conditions.[18] When a patient develops cardiac arrest temporally related to exercise, the physician should check to see if beta-blockers had been prescribed.

The magnitude of hyperkalemia with exercise is attenuated in well-conditioned athletes.[14] Conditioning enhances Na^+-K^+-ATPase activity, probably allowing more rapid reuptake of released potassium.

Hyponatremia

Because loss of sweat elevates plasma osmolality and serum sodium concentration, it is surprising that cases of hyponatremia have been reported with exercise. Since sweat is hypotonic, the only way that hyponatremia can develop is if the athlete consumes large amounts of water.[17] Even in this case, the kidneys should still be able to protect against hyponatremia by producing a dilute urine. Something, therefore, interferes with the kidney's powerful diluting capacity in these patients. One possible mechanism could be inappropriate antidiuretic hormone secretion (common in conditions of severe pain, perhaps an indirect effect of exercise).

Since the hyponatremia occurring with exercise develops acutely, it may be associated with severe symptoms such as seizure, change in mental status, or coma. When these symptoms occur, emergency treatment with hypertonic saline is necessary. In milder cases, intravenous sodium chloride is usually adequate.

Prevention Hyponatremia can only develop when water ingestion exceeds insensible losses. As long as water intake is less than or equal to insensible losses via sweat, serum sodium concentration should not change significantly. Therefore, replacement fluids should be consumed in quantities that do not greatly exceed losses. If volume losses were replaced with hypotonic electrolyte-containing fluids, the small risk of hyponatremia would be nearly eliminated.

Lactic Acidosis

Lactic acidosis may occur after exercise in association with heatstroke or as an isolated condition. The symptoms seen with lactic acidosis are identical to the symptom complex found in any form of metabolic acidosis. Dyspnea is common as the patient hyperventilates to compensate for the acidosis. Neurologic symptoms occur only rarely, causing lethargy or frank coma. The most important presentation, but also one that is very rare with exercise, is that of cardiac arrest: arrhythmias are common when blood pH falls below 7.1.

The etiology of exercise-induced lactic acidosis is related to an imbalance in the supply and demand for oxygen in muscle cells. The most efficient conversion of glucose to adenosine triphosphate (ATP) occurs by complete passage from glycolysis through the Krebs cycle and on through oxidative phosphorylation. When oxygen is not available, as in muscle cells during intense exercise, pyruvate, the end product of glycolysis, is converted into lactic acid. The acid produced is rapidly buffered in plasma, and the lactate is converted back into pyruvate in the liver and kidneys.

During intense exercise, lactate levels in the blood can rapidly increase twentyfold.[15] In extreme exercise, lactic acid is produced in such great quantities that it cannot be effectively buffered,[18] systemic pH may transiently fall to as low as 6.8 under these conditions.[18] Lactate is rapidly metabolized by the liver during rest, correcting the systemic pH to normal. It is very rare for acidosis occurring during exercise to cause any clinically important problems in healthy individuals. When acidosis persists after rest, hepatic dysfunction, with resultant decreased metabolism of lactate, should be suspected.

Acute Renal Failure

An uncommon but quite dangerous complication of exercise is acute renal failure. The pathogenic mechanisms relate to the loss of fluids and electrolytes as discussed above. This condition should be suspected in a patient who has recently performed strenuous exercise, has decreased urine output, or has symptoms of uremia.

The mechanism of acute renal failure in this setting probably involves the combined effects of hypovolemia and rhabdomyolysis. The hypovolemia leads to renal ischemia and decreased flow through the renal tubules. This results in increased myoglobin concentrations within the tubules, which predispose the patient to acute tubular necrosis by mechanisms that are not completely understood.

Significant rhabdomyolysis as a result of exercise is uncommon. However, it is important that it be recognized early by the clinician, because immediate prophylactic measures may help prevent acute renal failure. Signs and symptoms of rhabdomyolysis include diffuse muscle pain, muscle swelling, hyperkalemia, hyperphosphatemia, hypocalcemia, and hyperuricemia. The patient's urine will be red, and the urinalysis will be positive for heme by dipstick, but examination of the sediment will not reveal red blood cells. Only demonstration of elevated serum levels of muscle enzymes, such as creatine phosphokinase, lactate dehydrogenase, and serum glutamic-oxaloacetic transaminase, will establish the diagnosis conclusively.

Exercise-associated rhabdomyolysis is usually related to severe muscle exertion. The astute clinician should keep in mind, however, other predisposing factors, such as hypophosphatemia, hypokalemia, alcohol or cocaine ingestion, certain medications such as lovastatin and gem-fibrozil, and viral infection.

When the diagnosis of rhabdomyolysis has been established, preventing renal failure is crucial. The primary measure should be administration of enough intravenous fluid to maintain urine flow rates of at least 200 mL/hr. Forced alkalization of the urine with infusions of bicarbonate and mannitol to maintain the urine pH greater than 6.5 may be effective.

The risk of developing acute renal failure related to exercise appears to be greatest after prolonged, strenuous exercise on hot days in underconditioned athletes.[20,21] Among military recruits, acute renal failure has been noted in the summer during the first few weeks of basic training.[21]

Prevention. Besides adequate replacement of fluids and electrolytes during and after exercise, other steps can be taken to avoid acute renal failure. First, people embarking on a new exercise program should increase the intensity of their workouts gradually. Second, athletes traveling to a climate hotter than that in which they usually exercise must realize that they are not fully heat acclimated. They must, therefore, initially temper the intensity of their exercise. Finally, on very hot days, anyone, regardless of their level of conditioning or heat acclimation, must exercise with care.

Abnormalities Found on Urinalysis

In the conditions described above, the extreme metabolic stress of exercise on muscle leads indirectly to renal dysfunction. In contrast, direct and hormonal effects on the kidneys are probably more important in the pathogenesis of hematuria and proteinuria.

Hematuria. Either gross or microscopic hematuria can occur with exercise. The prevalence of microscopic hematuria depends on the type of activity, ranging from 55% in football to 80% in swimming or track and field.[22] One should keep in mind that red urine does not necessarily indicate hematuria, because myoglobinuria and hemoglobinuria also cause the urine to appear red. Red blood cells in the urine sediment confirm the diagnosis of hematuria.

The mechanism of hematuria occurring after exercise varies. With contact sports such as football, direct trauma to the kidneys is likely an important factor. With long distance running, it has been suggested that trauma to the bladder from the repeated up and down motion might be important.[23] Others[24] have found red cell casts and dysmorphic red blood cells in the urinary sediment, suggesting bleeding of glomerular origin. It is not known how exercise might cause glomerular bleeding.

Hematuria is a common presenting problem in clinical practice. When hematuria is due to exercise, the prognosis is excellent; the condition usually resolves within 48 hours. It is crucial, therefore, to inquire about recent strenuous exercise in the initial evaluation. When exercise is suspected to be the cause of the hematuria, a urinalysis should be repeated 48 to 72 hours after a period of rest; if the urine is free of blood at that point, no further workup is indicated. If the hematuria persists, a full evaluation as indicated by the patient's age should be initiated.

Proteinuria. Urinary excretion of protein is increased after strenuous exercise in 70% to 80% of subjects.[25] Dipstick readings for protein are found to be in the range of 2+ to 3+, with maximal excretion in the 30 minutes following exercise.[22] Twenty-four-hour collections of urine after exercise contain up to 300 mg of protein (normal<150 mg).

The mechanism of proteinuria occurring after exercise is probably related to a decrease in plasma volume as discussed earlier. This leads to increased renin and angiotensin II secretion,[26] with a resultant increase in the glomerular filtration fraction. With the increased filtration of water, the concentration of plasma proteins in glomerular capillaries rises, causing increased movement of protein across the glomerular basement membrane by means of diffusion and convection.[27]

Because proteinuria is a common clinical finding, the initial evaluation of any patient with proteinuria begins with a consideration of factors that may cause transient proteinuria, such as exercise. (Other causes include congestive heart failure and fever.) As with hematuria, proteinuria should always abate within 48 hours, and the prognosis in such cases is excellent. When proteinuria persists beyond this point, an evaluation for an underlying renal disease is necessary.

An Exercise in Prevention

When an active patient appears to have a renal illness or injury, it's important to elicit a history of recent strenuous exercise. This information is essential for proper diagnosis and expedient treatment. In addition, sports physicians need to stress to their patients the role of adequate fluid and volume replacement in both preventing and immediately correcting renal and electrolyte complications. ❑

References

1. Better OS: Impaired fluid and electrolyte balance in hot climates. Kidney Intl Suppl. 1987, 21:S97-S101
2. Costill D. L., Cote R. Fink W: Muscle water and electrolytes following varied levels of dehydration in man. J. Appl Physiol 1976:40(1):6-11
3. Rowell LB, Brengelmann GL, Blackmon JR, et al: Splanchnic blood flow and metabolism in heatstressed man. J Appl Physiol 1968:24(4):475-484
4. Knochel JP, Beisel WR, Herndon EG Jr, et al: The renal, cardiovascular, hematologic and serum electrolyte abnormalities of heatstroke. AM J Med 1961:30:299-309
5. Austin MG, Berry JW: Observations on one hundred cases of heatstroke. JAMA 1956:161(16): 1525-1529
6. Coyle EF, Montain SJ: Benefits of fluid replacement with carbohydrate during exercise. Med Sci Sports Exerc 1992:24 (9 suppl):S324-S330.
7. Noakes TD: Fluid replacement during exercise. Exerc Sport Sci Rev 1993:21:297-330
8. Rosa RM, Silva P, Young JB, et al: Adrenergic modulation of extrarenal potassium disposal. N Engl J Med 1980:302(8):431-434
9. Struthers AD, Quigley C, Brown MJ: Rapid changes in plasma potassium during a game of squash. Clin Sci 1988:74(4):397-401
10. Rose LI, Carroll DR, Lowe SL, et al: Serum electrolyte changes after marathon running. J Appl Physiol 1970: 29(4):449-451
11. Coester N., Elliott JC, Luft UC: Plasma electrolytes, pH, and EGG during and after exhaustive exercise, J Appl Physiol 1973:34(5): 677-682
12. Ledingham IM, MacVicar S, Watt I, et al: Early resuscitation after marathon collapse, letter, Lancet 1982: 2(8307):1096-1097
13. Clausen T, Wang P., Orskov H, et al: Hyperkalemic periodic paralysis: relationships between changes in plasma water, electrolytes, insulin and catecholamines during attacks. Scand J Clin Lab Invest 1980:40(3):211-220
14. Knochel JP, Blanchley JD, Johnson JH, et al: Muscle cell electrical hyperpolarization and reduced exercise hyperkalemia in physically conditioned dogs. I Clin Invest 1985:75(2): 740-745
15. Lindinger MI, Heigenhauser GI, McKelvie RS, et al: Blood ion regulation during repeated maximal exercise and recovery in humans. Am J Physiol 1992:262(1 pt 2):R126-R136
16. Lim M, Linton RA, Wolff CB, et al: Propranolol, exercise, and arterial plasma potassium, letter, Lancet 1981:2(8246):591
17. Irving RA, Noakes TD, Buck R, et al: Evaluation of renal function and fluid homeostasis during recovery from exercise-induced hyponatremia. I Appl Physiol 1991:70(1):342-348
18. Osnes JP, Hermansen L: Acid-base balance after maximal exercise of short duration. J Appl Physiol 1972:32(1):59-63
19. Better OS, Stein JH: Early management of shock and prophylaxis of acute renal failure in traumatic rhabdomyolysis, N Engl J Med 1990:322(12):825-829
20. Schrier RW, Henderson HS, Tisher CC, et al: Nephropathy associated with heat stress and exercise. Ann Intern Med 1967:67(2): 356-376
21. Vertel RM, Knochel JP: Acute renal failure due to heat injury: an analysis of ten cases associated with a high incidence of myoglobinuria. Am J Med 1967: 43(3):435-451
22. Cianflocco AJ: Renal complications of exercise. Clin Sports Med 1992:11(2): 437-451
23. Kallmeyer JC, Miller NM: Urinary changes in ultra long-distance marathon runners. Nephron 1993: 64(1):119-121
24. Fassett RG, Owen JE, Fairley J, et al: Urinary red-cell morphology during exercise. Br Med J (Clin Res) 1982:285(6353):1455-1457

25. Alyea EP, Parish HH Jr: Renal response to exercise-urinary findings. JAM 1958(7):167:807-813

26. Costill DL, Branam G, Fink W, et al: Exercise induced sodium conservation: changes in plasma renin and aldosterone. Med Sci Sports 1976:8(4):209-213

27. Goldszer RC, Siegel AJ: Renal abnormalities during exercise, in Strauss RH (ed): Sports Medicine, ed 2. Philadelphia, WB Saunders Co, 1991, p. 156

****

Questions:

1. What are some of the ramifications to the body that occur routinely with physical activity?

2. How much sweat can be lost from the body during strenuous exercise on a hot day?

3. What are the primary ingredients found in the sports drinks recommended to replace water and electrolyte losses?

Answers are at the end of the book.

Part Six:

Health and the Environment

Public interest is growing about the deleterious effects of estrogenlike pollutants in the environment. Scientists are focusing their studies on a new hormone-mimicking contaminant called bisphenol-A (BPA) which has been found, of all places, in the plastic linings of canned goods. During the sterilization process, BPA can leach out of the plastic and into the diets of people who consume them. Although BPA is not as potent as estradiol, its ability to persist in the body without causing adverse effects is unknown.

Additional Source of Dietary 'Estrogens'

by J. Raloff

Many canned foods on supermarket shelves contain small quantities of an estrogen-like pollutant, a new study reports. This hormone-mimicking contaminant — bisphenol-A (BPA) — appears to leach from the plastic resins coating the inside of affected cans.

Exposure to estrogen mimics has become a source of growing concern since recent studies began linking these ubiquitous contaminants with increased risks of breast cancer and reproductive abnormalities. During the past 4 years, endocrinologists have identified two types of plastics that can shed estrogen-like constituents.

Realizing that many food processors coat cans to avoid flavor-altering chemical reactions between the cans and their contents, Nicolás Olea and his coworkers at the University of Granada in Spain analyzed 20 different brands of canned goods. Purchased locally and in the United States, these included corn, artichoke hearts, mushrooms, tomatoes, and peas.

BPA turned up in roughly half of the items, the researchers report in the June ENVIRONMENTAL HEALTH PERSPECTIVES. Food processors note that about 40 percent of food cans in Spain are lined with plastic, compared to 85 percent in the United States.

Two years ago, David Feldman and Aruna V. Krishnan of Stanford University School of Medicine reported that BPA can leach from plastic subjected to high temperatures, such as those that occur during the autoclaving of laboratory equipment. Olea told SCIENCE NEWS that the sterilization of canned foods closely resembles that process.

Once plastic has been heated, BPA can continue to leach out. For instance, when a plastic-lined can was washed out and refilled with water, that water soon picked up measurable quantities of BPA, Olea's team reports. The pH of the food did not appear to affect leaching.

Where present, BPA occurred in trace quantities — just 4 to 22 micrograms per 300 grams of food. That's well below the 3 milligrams per kilogram of BPA allowed under regulations set by the European Union. BPA is also FDA-approved, and "no research or experience has suggested it might cause any adverse effects," says Roger Coleman of the National Food Processors Association in Washington, D.C.

Feldman's studies indicate that BPA possesses only one-thousandth the potency of estradiol, the major estrogen in humans. However, the body breaks down estradiol quickly, notes endocrinologist Ana M. Soto of Tufts University School of Medicine in Boston. If BPA lasts longer than estradiol or if the body cannot inactivate BPA as efficiently, "then it might prove more active than it at first sight appeared," she says. ❑

****

Questions:

1. What are the associated risks of exposure to estrogenlike pollutants?
2. How does BPA get into the environment?
3. How potent is BPA?

Answers are at the end of the book.

The pesticide DDT has been banned for more than 20 years in most Western countries because of its persistence and accumulation in the environment. Over time we learned that p,p'-DDE, the main metabolite of DDT was an environmental disaster as it caused reproductive abnormalities in wildlife. It has now been shown to be a potent anti-androgen, reeking havoc on the male reproductive tract. This information is raising suspicions in the health field in light of the decline in average sperm counts and increase of testicular cancer in normal men. DDT, as well as other 'hormonally active' compounds, still pose a health threat as they can be carried in from developing countries in food and animal products. If these trends are true, it means history is not ready to close the book on DDT.

Another DDT Connection

by Richard M. Sharpe

As we celebrate the half-centenary of the end of the Second World War in Europe, it will come as a surprise to many readers that the pesticide DDT played a part in the Allied victory. At the time, DDT was hailed as the new war weapon of the Allies[1] because of its ability to kill the lice responsible for the spread of typhoid. Only 20 or so years later was it realized that DDT was environmentally disastrous, as portrayed in Rachel Carson's *Silent Spring*. Now, 50 years on, comes another twist to the story, with the discovery by Kelce and colleagues that *p.p'*-DDE, the main metabolite of DDT in the body, is a potent anti-androgen. The fact that this discovery comes at a time of increasing concern about adverse changes in male reproductive health[3,4] is a remarkable coincidence. Or is it more than this?

There is evidence of a substantial fall in average sperm counts in normal men[5] and a progressive increase in incidence of testicular cancer[6] over the past half-century. There is other, although more equivocal, evidence that abnormal development of the penis (hypospadias) and mal-descent of the testes (cryptorchidism) have also increased in incidence.[4] It has been hypothesized that all of these changes are interrelated and that they may have a common origin in fetal and/or neonatal life.[3] But the veracity of the data on falling sperm counts continues to be disputed,[7] although a recent study of 1,300 fertile semen donors from the Paris region showed that sperm counts have dropped on average by around 2 per cent each year over the past 20 years.[8] In Denmark, where the lifetime risk of a male developing testicular cancer is now nearly 1 per cent, the government has been sufficiently concerned to commission a detailed report,[4] the findings of which are to be discussed in the Danish parliament in the autumn.

If these trends are true (and the increase in incidence of testicular cancer is beyond dispute), the cause (or causes) must lie in changes in our environment or lifestyle over the past 50 years. The most strongly argued case has been that human exposure to a range of weakly oestrogenic chemicals, which are ubiquitous in the environment, is to blame. This case is based, first, on evidence that exposure of man and animals to inappropriate levels of oestrogens during fetal and neonatal life will result in a spectrum of male reproductive disorders similar to those listed above.[3,4] Second, all these oestrogenic chemicals have been introduced into the environment in large amounts during the past 40-60 years.[4] They include certain detergent breakdown products and various plastics additives and stabilizers. However, it has long been known that one isomer of DDT (*o,p'*-DDT) and related chlorinated pesticides can also exert oestrogenic effects (so too can some modern pesticides).[4] Oestrogenic ('feminizing') and anti-androgenic ('demasculinizing') effects often manifest themselves in the same way, but work through oestrogen and androgen receptors, respectively. The discovery that *p,p'*-DDE is a potent antiandrogen means that 'DDT' has the capacity to interact with both receptors. As

Reprinted by permission from the author, Richard M. Sharpe and *Nature,* Vol. 375, pp. 538-539.

androgens have a key role in development of the male reproductive tract and in masculinization of the body in the male during fetal life, Kelce *et al.*[2] rightly ask whether *p,p'*-DDE could be responsible for some or all of the reported adverse trends in male reproductive health.

Use of DDT has been banned or restricted in most Western countries for 20 or more years, largely because of its persistence (it has a half-life of about 100 years) and its accumulation and concentration in food chains and because of a range of deleterious effects on reproduction in wildlife, DDT is still detectable in us all (mainly in body fat), despite the ban on its use. It is not, however, this present burden of DDT that is the centre of concern; rather it is the level of exposure in the period from the 1940s to the 1960s when it was still in use. Could it have exerted effects then on developing males which have only become apparent in more recent years? To address this possibility we need to know the level of exposure of the general population to DDT (DDE) when it was in use and whether such exposure would be sufficient to cause reproductive defects.

When first introduced, DDT was used in ways that would be unacceptable now. For example, when Naples was liberated by the Allies in the war, the whole population was 'dusted' with DDT.[1] Allied soldiers were similarly treated at recruitment and were issued with DDT-impregnated shirts when going overseas. In the tropics DDT was used in mosquito control and in many countries (including Britain) it was used in household flysprays. Today we would not contemplate using one of the safer modern pesticides in many of these ways, let alone a persistent chemical such as DDT. We can only guess what levels of exposure to DDT this led to in the general population, but measurements in dwellings in which DDT was used suggest that this exposure would probably be sufficient to exert significant antiandrogenic effects.[2] Male alligators in Lake Apopka in Florida contain high levels of *p,p'*-DDE and have abnormally small penises, amongst other reproductive changes.[9] Administration of pharmaceutical antiandrogens (used in the treatment of prostatic cancer) to animals during fetal and neonatal life can result in a range of reproductive tract disorders (small penis, hypospadias, cryptorchidism),[10] although sperm counts and testicular germ cell cancer in adulthood have not been studied.

If we are currently witnessing the legacy of DDT use in the past, can we at least comfort ourselves with the knowledge that things can only get better? Unfortunately not. Present use of DDT in developing countries (mainly for malarial control) probably exceeds the level of its use historically. Mexico and Brazil each used nearly 1,000 tonnes of DDT in 1992 according to World Health Organization figures.[11] As well as posing a health threat in these countries, the export of DDT through animals, crops and the atmosphere to the countries where its use is now restricted remains likely. And we still have the matter of environmental oestrogens on our doorstep to clear up.

Perhaps the most remarkable aspect of the findings of Kelce *et al.*[2] is that it has taken 50 years to discover that the main metabolite of DDT is an anti-androgen. This, coming on top of the recent discovery of oestrogenicity of several abundant environmental chemicals, must mean that our current methods of testing and surveillance of manufactured chemicals require modification so that the release of 'hormonally active' compounds to the environment can be minimized. For the present, we do not know whether such chemicals pose a significant risk to man, because we lack accurate data on the levels and routes of human exposure and on whether, and at what levels, adverse reproductive changes can be induced (in animals). Collecting this information is therefore a priority. Only then will we be in a position to assess whether these chemical skeletons in our cupboards are likely to have contributed to male reproductive disorders. ❑

References

1. *The Guardian,* 2 August 1944; reprinted 2 August 1994.
2. Kelce, W. R. *et al, Nature* **375**, 581-585 (1995).
3. Sharpe, R. M. & Skakkebæk, N. -E. *Lancet* **341**, 1392-1395 (1993)
4. Toppari, J. *et al, Male Reproductive Health and Environmental Chemicals with Estrogenic Effects* (Report No. 290 for the Danish Environmental Protection Agency, 1-166, 1995).

5. Carlson, E., Giwercman, A., Keiding, N. & Skakkebæk, N. -E. *Br. Med. J.* **305**, 609-612 (1992)
6. Adami, H. *et al, Int. J. Cancer* **59**, 33-38 (1994).
7. Bromwich, P., Cohen, J., Stewart, I. & Walker, A. *Br. Med. J.* **309**, 19-22 (1994).
8. Auger, J., Kuntsmann, J. M. Czyglik, F. & Jouannet, P. *New Engl. J. Med.* **332**, 281-285 (1995).
9. Guillette, L. J. Jr *et al, Environ, Health Perspect,* **102**, 680-688 (1994).

****

10. Imperato-McGinley, J., Sanchez, R. S., Spencer, J. R., Yee, B. & Vaughan, E. D. *Endocrinology* **131**, 1149-1156 (1992).
11. PAHO (Pan American Health Organization) *Status of Malaria Programs in the Americas* 40th Report (World Health Organization, Washington, DC, 1992).

Questions:

1. What role do the androgens play in the body?

2. Why was DDT banned?

3. What is the half-life of DDT?

Answers are at the end of the book.

Hormone-mimicking pollutants can open a Pandora's box of health risks. Long-term exposure to estrogenic-like compounds may be linked to the increase in breast cancer and infertility in men. The presence of DDT in the environment has sparked interest recently due to some scientific evidence stating that estrogenic chemicals act through nonestrogenic pathways by blocking the activity of male sex hormones. It is DDE, a metabolite of DDT, which can bind the androgen receptor and trigger emasculating endocrine changes in animals. What impact these pollutants have during human development and the individual receptor system targets are questions which must be addressed to better understand environmental hormones.

Beyond Estrogens

Why unmasking hormone-mimicking pollutants proves so challenging

by Janet Raloff

Atomic representations of three common PCBs. EPA's new computer models of androgens and estrogens correctly predicted that the top two would have little or no affinity for sex hormone receptors but that the third PCB would.

Over the past year, the news media have hammered home the message that most of the animal kingdom is bathed in a sea of pollutants that act like estrogen, the primary female sex hormone. Reinforced by photos of birds, fish, and alligators exhibiting gonadal malformations, a growing awareness and acceptance of these risks has crept into the public's consciousness.

Epidemiologists, endocrinologists, and reproductive biologists have suggested that long-term exposure to these increasingly pervasive pseudoestrogens might underlie an apparent breast cancer epidemic in women and fertility impairments in men.

But follow-up studies now indicate that the initial message was too simplistic.

Estrogenic chemicals do pose risks to people and wildlife by binding to and unlocking the genetically programmed activity of estrogen receptors dispersed throughout the body. But scientists at the Environmental Protection Agency's Health Effects Research Laboratory in Research Triangle Park, NC, have now identified a host of pollutants that functionally mimic estrogens in animals yet act through a nonestrogenic pathway: They block the activity of male sex hormones.

Related rodent studies at this laboratory and elsewhere have demonstrated that another class of pollutants — one that includes TCDD and other dioxin-like chemicals — triggers emasculating endocrine changes through yet another mechanism.

Finally, a series of studies now under way at the University of West Florida in Pensacola is looking at the *masculinizing* effects of paper mill wastes on female fish. Ichthyologist Stephen A. Bortone found he could trigger gender-bending changes similar to those noted in the wild by exposing laboratory fish to any of five androgenic substances, including testosterone, the primary male sex hormone. Initially, this led him to assume that the water pollutant affecting fish downstream from paper mills must be some kind of environmental androgen. Now, he concedes it might be an estrogen blocker. Tests over the next year or so may help resolve the issue.

These new findings reinforce the pivotal role of animal studies in identifying pollutants with a hormonal alter ego, argues University of Missouri biologist Frederick S. vom Saal. No battery of biochemical, cellular, or tissue-culture tests exists — or is likely to come along soon — that can fully map the myriad subtle changes in development that early exposures to gender-bending pollutants may cause, he maintains. It's even less likely that scientists can predict how those changes will play out in terms of disease or fertility, he notes.

While these new data muddy the waters, they do nothing to downplay toxicological concerns associated with the pollutants. If anything, they expand the list of suspected crimes, the cast of potential troublemakers, and the sites of their potential villainy.

Over the past 3 years, studies at the University of Florida in Gainesville have illustrated some of the important trends in this developing field.

Scientists there have been working to ferret out the mechanisms by which spills of a DDT-laced pesticide along the shores of Lake Apopka poisoned local alligators. Rates of hatching and hatchling survival remain extremely low at this, the state's fourth largest body of freshwater. The young that survive possess grossly impaired reproductive systems.

The first clue that the endemic pesticide contamination there might have hormonal consequences came from data on the animals' sex hormones. The proportion of estrogen was much higher than it should be, largely owing to a near absence of testosterone — even in males.

The fact that mice treated with the estrogen-mimicking drug diethylstilbestrol (DES) exhibited ovarian abnormalities resembling those in Apopka's female alligators "initially argued that these [reptiles] had been exposed to an estrogen," recalls Gainesville biologist Louis J. Guillette, Jr.

But on further reflection, he decided, excess estrogen should also increase penis size in males — the opposite of what he observed in juveniles at Apopka. He says their smaller-than-normal penises suggested a deprivation of androgens, or male sex hormones. "So I started playing with the idea that the phallus phenomenon, at least, was due to an antiandrogenic effect of DDE," the long-lasting and most toxic, breakdown product of the pesticide DDT.

Last year, Guillette brought up this antiandrogen hypothesis during a presentation of his work at a meeting in Atlanta; among the attendees was L. Earl Gray Jr., a reproductive toxicologist with EPA.

Unknown to Guillette, Gray and William R. Kelce, also from EPA's Research Triangle Park laboratory, had already come to much the same conclusion. "The effects that Lou was seeing in the male alligators looked almost identical to the developmental effects we were seeing with vinclozolin in our rats," Kelce explains.

He and Gray were part of a research team that had just identified this widely used fungicide as the first known androgen blocker in the environment. By binding to androgen receptors in rodents exposed before birth, vinclozolin prevents sufficient testosterone from reaching androgen receptors.

The legacy of this early exposure in males included testes that failed to descend, infertility, a partially unfused phallus, and the development of a "vaginal pouch" characteristic of the female reproductive system. In other words the males had suffered not only a demasculinization, but also a feminization.

Since DDE was the only major pollutant isolated from the eggs and tissues of Apopka alligators, Kelce and Gray set out to investigate its ability to bind to androgen receptors.

When Guillette stopped by their laboratory a month or so after his Atlanta talk, he found Gray and Kelce "looking like they'd swallowed canaries." The reason for their satisfied smiles, he quickly learned, was that their work had just confirmed that DDE readily binds to the androgen receptor.

Kelce, Gray, and their coworkers shared these and follow-up data with the rest of the world in the June 15 NATURE. Not only does DDE fail to unlock this receptor's normal sex steroid activities, but by tying up the receptor it also prevents true androgens from triggering their intended effects.

The roughly 60 parts per billion (ppb) concentration of DDE needed to block androgen receptors falls well below that seen in some DDT-contaminated communities. Eggs of some feminized male Apopka alligators contained up to 5,800 ppb. And blood from South Americans whose homes were treated with DDT to control malaria carries as much as 140 ppb.

Moreover, this DDT metabolite can cross the placenta. Tissues from stillborn infants in Atlanta during the mid-1960s — priors to the 1972 phaseout of DDT in the United States — contained DDE concentrations as high as 650 ppb in the brain, 850 ppb in the lung, 2,740 ppb in the heart, and 3,570 ppb in the kidney, Kelce's team notes.

Though most industrialized Western countries banned DDT use 2 decades ago, the toxicant still turns up in foods and soil throughout those nations. Why? In part, because other countries continue to use the compound. Indeed, points out Richard M. Sharpe of the Medical Research

Council's Reproductive Biology Unit in Edinburgh in the same issue of NATURE, "Present use of DDT in developing countries (mainly for malarial control) probably exceeds the level of its use historically. Mexico and Brazil each used nearly 1,000 tons of DDT in 1992."

Kelce's team concludes that "the reported increased incidence of developmental male reproductive system abnormalities in wildlife and humans may reflect antiandrogenic activity of the persistent DDT metabolite DDE."

Three months earlier, the same EPA researchers reported new data showing that prenatal exposure to TCDD, the most toxic dioxin, permanently alters reproductive function in male rats and hamsters. The doses used were relatively small, they point out in the March TOXICOLOGY AND APPLIED PHARMACOLOGY — well below those causing observable changes in the mothers.

Unlike vinclozolin, TCDD did not feminize exposed animals. Instead, males of both species suffered a slight emasculation of their reproductive system and sexual behavior. Though subtle, these changes did translate into "fairly large effects on the ejaculated sperm counts — even though the testes were only minimally affected," Gray told SCIENCE NEWS. He suspects the early exposures altered the development of the males' epididymis — a gonadal structure that stores sperm and plays a role in ejaculation — and possibly that of related sex accessory glands.

Exposed male hamsters also developed mild epididymal inflammation, which Gray says "might imply a slight structural change." Though hamsters are "typically insensitive to the lethal effects of dioxin," he notes, "we've now shown the hamster fetus is not insensitive."

This suggests that adult and fetal toxicology "do not correlate," he says. "So if humans are insensitive to the lethal effects of dioxin, that might not mean the fetus will be also."

"We still believe that dioxin is acting as a potent environmental hormone," observes Linda Birnbaum, director of environmental toxicology at EPA's Research Triangle Park facility. TCDD exerts its gender-bending effects by binding to its own distinct (aryl hydrocarbon) receptor. Once that binding occurs, the dioxin receptor complex "sends out mixed signals that alter the expression of a multitude of different hormones or hormone receptors," she says.

Under some circumstances, TCDD apparently produces antiestrogenic effects. At other times, this dioxin appears to need the presence of estrogen in order to unleash its toxicity. In still other cases, TCDD's effects resemble those of estrogen.

"We know dioxin doesn't bind to the estrogen receptor," Birnbaum says. What it appears to do "is affect the expression of that receptor."

Indeed, vom Saal maintains, "it's virtually impossible to disturb any part of the endocrine system without seeing responses in other parts." So adding hormone mimics or blockers to any part of the grand, interrelated community of hormones at work in the body may unexpectedly evoke changes elsewhere.

At no time is the body more susceptible than during fetal development.

Each individual begins life as a single fertilized cell that will divide again and again — eventually differentiating into tissues as diverse as muscle, bone, and brain. Yet each of these cells still carries the same genes. The difference, vom Saal explains, "is that during development, some genes are masked, others turned on. Hormones do that."

By tweaking the system as tissues are differentiating, he says, "you may not damage the genes — just turn them off forever or turn them on at abnormal rates or times."

What this means, Guillette explains, is that the hormones' role in organizing fetal development does not stop at making sure organs end up on the right place. Hormones "even direct organization on the cellular and genetic level," he says, sometimes modifying receptors or even changing the number present. This is why adult male rodents, exposed to DES as fetuses, respond to estrogen stimulation by producing reproductive proteins ordinarily seen only in females, he says. "Their cells were reorganized so that they now respond inappropriately."

These features help explain why it's so difficult to identify environmental hormones. "We now know that no one technique, assessment, or species will be able to tell us whether an ecosystem is polluted," Guillette says.

In fact, vom Saal adds, "we have not even begun in science to identify the multitude of hormones active during development and the receptor systems [that may be vulnerable]." In that sense, researchers don't yet know where to look for toxicity — beyond the classic, albeit simplistic, approach of identifying agents that bind to either estrogen or androgen receptors.

Moreover, vom Saal charges, these new data from studies on environmental hormones suggest that even the standard animal models for evaluating a chemical's safety — high doses over short periods — are inadequate. When it comes to hormones, he explains, "the body has a protective system. Whenever hormone levels get too high, the body basically shuts off its receptors [for those hormones]."

"So to find out how chemicals really impact development, you must know at what levels they occur in the environment and then go out and test animals at that level. Surprisingly," he says, "when we do that, we find effects not predicted by the very high dose studies."

In the future, vom Saal would like to see every licensed chemical — and every compound proposed for commercial introduction — tested in developing animals at concentrations likely to be encountered in the environment.

The concept is daunting. Where would toxicologists begin?

Research Triangle Park chemist Chris L. Waller hopes to offer some concrete suggestions soon. He leads a team working on computer models of the features that define estrogens, androgens, and other important hormones.

Waller's group hopes to assemble a family of tools that can be used to examine a molecule's three-dimensional structure and the electric charge associated with it; from these data, the team will predict the molecule's potential for binding with particular receptors. But because no one has yet identified the structure of those receptors, Waller's group is stuck with having to look for similarities between known members of each class of hormones (and their mimics). Among the compounds they've used in building the androgen model are about 20 unpublished antiandrogens — some as potent as DDE — that Kelce's group identified through more intuitive means.

"The androgen model I've put together is very predictive," Waller says. The group has refined the estrogen receptor model fairly well and has begun a model to screen for analogs of progesterone, another female sex hormone.

Kelce expects these models eventually to run through a library of thousands of structures in just a few hours. Compounds flagged as suspect could then be evaluated further in test-tube or animal toxicity assays.

In April, EPA convened a small international scientific conference to discuss the possible need for a U.S. research strategy to better understand environmental hormones. "The consensus of the workshop was that . . . [there] was sufficient concern to warrant a concerted research effort"— one that places especially high priority on identifying the risks of these compounds to the developing reproductive system, says organizer Robert Kavlock, director of developmental toxicology at the agent's Research Triangle Park facility.

Based in part on the conference, he says, EPA plans to beef up funding for research on environmental hormones, including their effects on wildlife and human health. Another by-product of the meeting, he says, will be the formation of a committee this month or next to coordinate new research efforts in this field by various government agencies, industry, and environmental groups. ❑

****

Questions:

1. How do estrogenic chemicals work through nonestrogenic pathways?

2. What is DDE's proposed antiandrogen mechanism?

3. Why are scientists concerned about DDE's effects during human development?

Answers are at the end of the book.

If you're in an exercise program or beginning to embark on one, its nice to know just how much exercise is adequate to produce the desirable health benefits associated with a longer life. Health officials from several organizations have been working on an exercise prescription that would be both beneficial and feasible for most healthy adults in the United States. According to recommendations, in order to receive the maximum benefits to your cardiovascular system, lung capacity and metabolic rate you need to do 20 minutes of aerobic exercise three times a week. Think of it as an investment in your future health.

How Much Exercise is Enough?

Harvard Women's Health Watch

If recent reports on the health benefits of physical activity are exercising your patience, it's little wonder. At the beginning of the year the prevailing prescription for fitness was one developed by the American College of Sports Medicine (ACSM) in 1985: 20 minutes of aerobic exercise activity intense enough to produce perspiration) three times a week. This recommendation was seemingly overturned in February when the ACSM and the Centers for Disease Control and Prevention (CDC) issued new, less stringent recommendations, advising all adults to perform a total of 30 minutes of moderate-intensity exercise (no sweat required) daily. Then, in April, analysis from the Harvard Alumni Health Study indicated that "moderate" exercise isn't enough; only vigorous (perspiration-producing) activity is associated with a longer life.

Although it may sound as though exercise experts keep changing their tune, they're really singing different songs. When scientists chosen by CDC and the ACSM got together last year, their mission was to go over the major studies linking exercise to health and longevity and to come up with a recommendation for a program that would be both beneficial and feasible for most healthy adults in the United States — the majority of whom are now sedentary. As such it was a compromise; it wasn't supposed to be the definitive exercise regimen, but a beneficial program that the greatest number of people might accept.

The Harvard researchers had a different purpose: to determine the level of exercise that was associated with the lowest number of deaths among 17,000 Harvard graduates who filled out lifestyle questionnaires in 1962 or 1966. They found that those who lived the longest were more likely to expend energy at a rate roughly equivalent to running at 6 mph for a total of 6 hours a week.

The Harvard results were surprising because, contrary to those of other studies, they indicated that moderate exercise had no benefits over a sedentary lifestyle.

It's worth noting that the subjects in the Harvard investigation, who enrolled in the university between 1916 and 1950, were all men. Although fewer exercise studies have been conducted in women than in men, most have indicated that moderate exercise has decided benefits.

What to do

You may want to think of the various exercise prescriptions presented recently as stepping stones to higher levels of fitness. The CDC/ACSM guidelines are a good place to begin. It isn't difficult to log 30 minutes of daily exercise by doing some errands on foot, parking the car or getting off the bus a few blocks from your destination, and taking the stairs instead of the elevator. Each minute you're on your feet can also contribute to the weight-bearing exercise necessary to preserve bone and reduce the risk of osteoporosis.

For conditioning your cardiovascular system, improving lung capacity, or raising your metabolic rate, you'll still have to work up a sweat; at least 20 minutes of aerobic exercise three times a week is probably necessary. And, if the Harvard findings can be extended to women, you may need to increase the number or duration of your workouts to increase your chances of living longer.

The hidden message

One aspect of observational studies is often lost in the telling: The active people weren't exercising because they were asked to but because they chose to, perhaps because they were in better shape to begin with. These days, as we try to come up with the right "dose" of physical activity, it's easy to forget that exercise isn't just good medicine — it is often great fun as well. ❑

****

Questions:

1. How did ACSM and CDC decide on the best exercise for most Americans?
2. How is aerobic exercise defined?
3. Why are weight-bearing exercises good?

Answers are at the end of the book.

-43-

Factor VIII is a blood-clotting protein well known to hemophiliacs who must inject it on a regular basis to survive. In order to have an ample supply of the protein, it must be extracted from donated blood or produced by genetically engineered bacteria. A new approach to this idea involves implantation of genetically altered cells into a patient that can produce factor VIII as needed. So far, reports show positive results with mice. In all likelihood, the same will occur for humans, making gene therapy for hemophilia a reality in the near future.

Fixing Hemophilia

by Sarah Richardson

Hemophilia is caused by a defect in a gene that codes for a crucial blood-clotting protein — for the one called factor VIII, in 85 percent of all cases. To avoid bleeding to death, hemophiliacs must regularly inject themselves with factor VIII that has been extracted from donated blood or produced by genetically engineered bacteria. A simpler and potentially cheaper solution, and one that would eliminate the need for regular shots, would be to give hemophiliacs the gene to make their own factor VIII. Researchers at Somatix Therapy Corporation in Alameda, California, have just taken an important step toward the goal of gene therapy. They have implanted in mice a clump of genetically altered cells that can churn out the missing protein.

Molecular biologist Varavani Dwarki and his colleagues first inserted the gene for factor VIII into a harmless retro-virus — a type of virus that splices its own genes into the DNA of a host cell. Then they used the virus to ferry the factor VIII gene into human skin cells. Although factor VIII is made primarily in the liver, skin cells are the most likely target for gene therapy because they are so easy to collect and to grow in culture.

But simply injecting genetically altered skin cells into a patient wouldn't work; the skin cells wouldn't survive in the bloodstream. So the researchers built a "pseudo-organ" consisting of alternating layers of skin cells and collagen, the fiber that forms connective tissue. The collagen pulled the 50 million or so skin cells into a tight, dime-size disk that was easy to implant and more likely to survive in the body.

As a first test of the pseudo-organ, Dwarki implanted it in mice, attaching it to the outside wall of the gut in a region rich with blood vessels. The cells survived a little over a week. (The mice did not reject the human tissue outright, because they belonged to a laboratory strain that lacks an immune system.) During that period they secreted far more factor VIII than a normal human needs, let alone a mouse.

The mouse experiment, says Dwarki, proves that gene therapy for hemophilia is possible in principle — that human cells can be engineered to produce factor VIII, and that those cells can be successfully implanted in a body. Dwarki's next goal is to implant genetically altered dog skin cells in hemophiliac dogs. If that works well, he says, "the day may come when we can grow a person's cells for a couple of weeks, then put in the gene, and then implant the altered cells. We hope this will provide a lifetime supply of factor VIII." ❑

****

Questions:

1. What happens to the blood without factor VIII?
2. How did molecular biologists insert the gene for factor VIII into human skin cells?
3. What are the advantages of gene therapy for hemophilia?

Answers are at the end of the book.

Am I at risk for contracting a dreaded disease such as AIDS or hepatitis through a blood transfusion? Many concerned patients are asking this question of hospitals and clinics based on a report by the Food and Drug Administration which says that in spite of screening, patients are at some risk for receiving tainted blood. Tell that to the 3.6 million people who get transfusions of whole blood or its products in a year. Health officials, blood banks and the federal government are evaluating the current practice at blood banks to determine if there is a way to improve screening without having to discard usable blood at an enormous cost. Without a consensus on the issue, the business of protecting our nation's blood supply remains nebulous.

Debugging Blood

Protecting people from tainted blood

by Tina Adler

Sally, the mother of a friend of mine, refused a transfusion after surgery. "I just don't trust the blood supply," she says. Sally felt weaker and her recovery took longer than if she had accepted fresh blood. But she says her peace of mind made the struggle worthwhile.

In increasing numbers, patients are avoiding the blood of strangers. They bank their own, ask friends to donate for them, or do without. Must they really go to such lengths, or has AIDS made them unduly afraid?

In spite of screening, about 1 in 225,000 units of blood reaching patients may harbor HIV, the AIDS-causing virus; 1 in 250,000 the hepatitis B virus; and 1 in 3,300 the hepatitis C virus, according to studies funded by the Food and Drug Administration. Those figures might provide some comfort if the diseases weren't so dreaded — AIDS has no cure, and hepatitis B and C can destroy the liver, causing death.

About 3.6 million people get transfusions of whole blood or blood products, such as platelets, each year; most require more than one unit, about a pint.

Last month, the National Institutes of Health in Bethesda, Md., asked a panel made up primarily of physicians to answer two questions. Should blood banks continue using three blood tests that screen for syphilis, non-A, non-B hepatitis, and — indirectly — AIDS? How can health officials improve their response to new organisms that may threaten the blood supply?

The panel's answers may well change the way blood banks do business.

The three tests the NIH panel reviewed prevent few cases of disease not prevented by prevented by other means, and they yield many false positive results — forcing blood banks to dump hundreds of thousands of units of good blood, the panel members assert.

False positives "not only contribute to the present blood shortage but also result in emotional, psychological, and financial costs to the donor," the panel explains in its report.

Donors who test positive for an infectious disease often undergo costly and complicated follow-up tests. Some repeatedly test positive, yet have no illness. And if their results reach their insurance companies, the insurers may try to deny benefits, panel members say.

To protect the public adequately, the group concludes, health officials must do much more than improve blood screening. No one agency or group has responsibility for telling blood blanks how to respond to a new disease, such as Chagas' disease. And no plan exists for preventing new diseases from threatening the blood supply.

People in the United States donate 12 million units of blood annually. The American Red Cross runs about half of the U.S. blood banks; hospitals and clinics — most affiliated with the American Association of Blood Banks (AABB) — operate the others. They all must screen blood donations for syphilis, hepatitis B and C, two types of HIV, two types of human T cell leukemia virus, and other infectious agents.

Most HIV-contaminated blood that reaches transfusion recipients comes from recently infected people who have not yet developed antibodies to the virus. HIV tests now in use detect only antibodies, which may take 25 days or more to show up, researchers say. Moreover, newly infected individuals are far more infectious than previously suspected.

Scientists have yet to find ways to kill or eradicate viruses and bacteria found in donated blood without also destroying the blood itself. So testing donated blood before giving it to patients remains the only protection available.

HIV researchers have developed tests that detect the virus at an earlier stage of infection than those the blood banks now use. But these more sensitive screens might not prove cost-effective if used widely, several investigators told panel members at a conference held last month at NIH.

In response to the dilemmas facing blood banks, the panel recommended nixing a test that measures the activity of the enzyme alanine aminotransferase (ALT) in the blood. It favored the continued use of the syphilis and the hepatitis B core antibody (anti-HBc) tests, which researchers developed to detect non-A, non-B hepatitis virus.

The FDA, which regulates blood collection, supports the NIH group's recommendations concerning the tests, an agency spokeswoman says. Officials of the AABB and the Red Cross are still reviewing the report.

ALT, a component of liver cells, enters the bloodstream in response to liver damage, such as that caused by hepatitis. However, heavy alcohol consumption, obesity, and, possibly, strenuous exercise — factors that do not make people unsuitable donors — also increase ALT activity, studies suggest.

Indeed, no clinical studies show that ALT screening improves the safety of blood transfusions, panel members conclude. Blood collection staffs now rely on more recently developed tests to detect hepatitis. The FDA does not require blood banks to use ALT, although all Red Cross centers still do and AABB recommends its use.

ALT tests cost little to manufacture but prove costly in other ways. Every year, blood banks discard roughly 200,000 units and turn away 150,000 potential donors because of elevated ALT readings. But few of those units actually pose a health risk, the panel asserts. In addition, physicians often recommend that their ALT-positive patients undergo further, more expensive liver tests, the panel reports.

Blood bankers discard about 20,000 units of blood each year as a result of false positives from the anti-HBc test, also commonly used in screening, says panel member Theresa L. Wright of the University of California, San Francisco. False positive anti-HBc results have caused health officials to reject "tens of thousands" of donors, the panel says.

Nonetheless, the test spots a small number of units infected with hepatitis B. People infected with hepatitis B may also have HIV, since both diseases can be spread through sexual contact or shared needles. So the anti-HBc indirectly spots HIV-tainted blood that would otherwise go undetected. These finds make up for the test's high false positive rate — at least until better HIV detectors come along, the panel concluded.

The test may catch, albeit indirectly, as many as one-third of the HIV-contaminated units that other screens miss, preventing roughly six cases of transfusion-transmitted HIV annually, according to panel member Jeffrey McCullough of the University of Minnesota Hospital in Minneapolis. However, the value of the anti-HBc test "is likely to decline with expanding HBV [hepatitis B virus] immunization," the NIH report warns.

Blood banks have tested blood for syphilis for more than 50 years, and transfusion-transmitted syphilis has become extremely rare. Whether the test deserves credit for that low infection rate remains uncertain, however.

It's unclear whether the syphilis test actually detects *Treponema pallidum,* the syphilis-causing bacterium, during its infectious stage, says McCullough. People who test positive for other diseases may also have syphilis, so other tests may deserve the credit for removing syphilis-infected blood, McCullough says.

Also, blood often gets refrigerated for more than 3 days, which destroys *T. pallidum.*

Syphilis screening excludes less than one HIV-positive donor annually whom other tests would have missed. Nevertheless, the panel concludes, blood bank staffs should continue to use the test until researchers determine exactly what role it plays in preventing transfusion-transmitted syphilis.

Tests now used only for research can spot HIV earlier than any of the anti-body-based methods available to blood banks. For example, one screen that looks for the p24 antigen, a protein on the surface of HIV, detects the virus 6 days sooner than the HIV tests blood banks use, Guillermo A. Herrera of the Centers for Disease Control and Prevention (CDC) in Atlanta told the conference. Yet blood banks may never use the p24 antigen test because it would cost a lot, yet save few additional lives, he and others assert.

If widely employed, the p24 screen could detect up to one-half of HIV-infected units of blood now going to patients from donors who haven't yet developed antibodies to HIV, says CDC's Lyle R. Petersen. Statistical modeling using the HIV test results of repeat donors indicates that roughly 35 such people donate each year, and existing tests pick up about one-fifth of them, he says.

What's more, half the patients who get transfusions die within a year from problems unrelated to their new blood. So the p24 test would help very few patients, Petersen points out. In addition, if clinics test for p24, more people at risk for HIV may donate blood to find out their HIV status, researchers say.

The NIH panel did not discuss p24 in its report because it didn't receive adequate information about the test, panel member Karen L. Lindsay of the University of Southern California in Los Angeles says.

Scientists can detect HIV's DNA and RNA in blood more quickly than they can find antibodies to HIV, Herrera says. However, these tests may not be cost-effective for blood banks either.

Someday, blood screening in general may prove less important. The federal government and biotechnology companies are spending hundreds of millions of dollars to find ways to eliminate or inactivate infectious agents in the blood, as well as to make artificial blood.

Baxter Healthcare Corp. in Round Lake, Ill., plans to seek permission from FDA this year to use an intravenous solution that can temporarily take on blood's oxygen-carrying responsibilities, Martha C. Farmer, director of product management for blood substitutes at Baxter, told SCIENCE NEWS. She thinks the new solution may replace blood in 15 to 20 percent of transfusions.

However, the technology for purifying the red blood cells, which most transfusion patients receive, "is nowhere near [human] trials," she says.

The guardians of the public's blood supply need to develop a strategy for protecting it from new infectious agents, panel members assert.

Part of that strategy should include getting more accurate medical histories from donors. That may serve as one of the best ways to keep bad blood from patients, they contend. Studies show that improvements made in the early 1980s to the questionnaires used by blood bank staff have helped reduce the transmission of HIV, hepatitis B, and hepatitis C.

Federal regulations require physicians to report cases of transfusion-transmitted diseases to CDC. However, no early warning system exists to alert blood banks to possible threats, says McCullough. No one group or agency has responsibility for monitoring what new agents may lurk in donors' blood or for determining what blood bank staff should do to keep such poisons out of their supplies, panel members say.

"There's no clear-cut communication system" for use by blood bankers and federal agencies responsible for protecting the public health, Lindsay says.

When a new organism appears, "the response isn't as coordinated as it could be . . . The federal system works slowly," Wright says. Adds Lindsay, "there's not an organized surveillance system."

These are discouraging words for anyone who may someday require someone else's blood — although they won't surprise my friend's mother or anyone else trying to get by without donor blood.

But health officials continue their campaign to convert the worrywarts: The risk of transfusion-contracted disease remains exceedingly low, they assert. ❑

****

Questions:

1. What tests are routinely done on donated blood?
2. How can infectious agents be removed from donated blood?
3. What other improvements can hospitals and blood banks make when screening blood donors?

Answers are at the end of the book.

Are women protected from coronary heart disease by estrogen? It does appear from studies that women suffer lower rates of heart attack when estrogen levels are high but more studies are needed to verify these observations. The female hormone has been shown to influence some of the factors that contribute to heart disease such as serum lipids, blood clotting and vessel dilation. By studying the connection between estrogen and the cardiovascular system, scientists hope to gain more knowledge about heart disease and how to best prevent it.

Estrogen and Your Arteries

Harvard Women's Health Watch

By now you're probably well aware that estrogen is considered to be responsible for our 15-20-year advantage over men in evading coronary artery disease. Not only do most women have little evidence of heart problems during their reproductive years, when estrogen levels are high, but those who take estrogen after menopause seem to have lower rates of heart attack.

Still, the evidence for estrogen's heart-sparing effects is mostly circumstantial. It has been derived, for the most part, from observational studies in which researchers looked at women who had suffered heart attacks or strokes and those who hadn't and determined how many in each group were taking or had taken estrogen. In these studies, although the women who had heart attacks were usually less likely to be taking estrogen, there was no conclusive proof that estrogen had protected the others. What remained unanswered was whether the protective effect was due to estrogen or whether the women on estrogen were merely more likely to see a doctor regularly and to have better health habits than those who didn't take hormones.

Randomized controlled trials like the Postmenopausal Estrogen/Progestins Intervention Trial (PEPI) and the Women's Health Initiative (WHI) are designed to put this question to rest. In the WHI, women with similar risk factors and medical histories are randomly assigned to take hormones or placebos and will be followed for several years to determine how many have heart attacks or strokes. If a higher percentage of the women on estrogen are spared, the hormone treatment is likely to have made the difference.

In the meantime, other studies have indicated that estrogen influences some of the factors that contribute to heart disease, including the following:

• *Serum lipids.* Because cholesterol is a major component of atherosclerotic plaque, serum lipid levels are important determinants of coronary-artery risk. The consensus among those who study blood lipids seems to be that for women, having a high level of high-density lipoprotein, or HDL, which helps to keep cholesterol from being deposited, offers even greater protection than having low total cholesterol levels — less than 200 mg/dL. In the PEPI trial, the women who took estrogen had significantly higher HDL levels than those on placebo.

However, estrogen's effects aren't entirely beneficial. In some studies it has been linked to increased levels of lipids called triglycerides, which are derived directly from the fatty acids in foods or are manufactured from carbohydrates within the body. There is growing evidence that high triglyceride levels may increase the risk of heart disease. For example, in the PEPI study, women who took estrogen had a greater increase in triglyceride levels than did those who took placebos. Whether the increase was high enough to affect heart-attack risk isn't known.

• *Blood clotting.* Both *thromboses* (clots that form in injured or narrowed arteries) and *emboli* (clots that break away and travel through the blood stream) can cause heart attacks and strokes. Although high doses of estrogen like those in early oral contraceptives have been linked to increased blood clotting, there is evidence that in lower doses estrogen may prevent clots from forming and even increase the body's ability to dissolve clots that do occur.

An increase of about 3% in levels of one clot-forming agent, fibrinogen, doubled the risk of heart attack among women in the Framingham Study, a large ongoing observational trial. In the PEPI study, fibrinogen levels rose by about 3% in women who were not on estrogen, but did not change among those who took it.

A recent report from the second generation of the Framingham Study indicated that restoring estrogen to premenopausal levels also improves clot-dissolving ability. Premenopausal women and postmenopausal women who were taking estrogen had lower levels of a substance called plasminogen activator inhibitor (PAI-1) than did postmenopausal women who were not taking estrogen. In this case "lower" is "better" because PAI-1 interferes with the natural clot-dissolving process.

• *Vessel dilation.* Scientists have discovered that estrogen is taken in by cells in both the muscular walls of blood vessels and the endothelium — the layer of tissue overlaying the muscle. In a small study of postmenopausal women conducted by investigators at the National Heart, Lung, and Blood Institute, estrogen injections that brought circulating estrogen to premenopausal levels led to dilated blood vessels. Laboratory studies have indicated that this effect may be due to nitric oxide or other vessel-dilators released by endothelial cells in response to estrogen. Some investigators think that estrogen may thus improve blood flow and prevent vessel spasms that contribute to heart attack and stroke.

• *The message.* Better knowledge of estrogen's effects on the cardiovascular system may provide more evidence in favor of hormone replacement therapy. At the same time, it may also enable researchers to develop treatments with similar effects for women who don't choose to take estrogen. ❑

****

Questions:

1. When are estrogen levels high during the life of a normal female?

2. How does estrogen affect serum lipids?

3. How does estrogen affect blood vessel dilation?

Answers are at the end of the book.

People in the United States are obsessed with losing weight. The $30 billion dollars spent each year in effort to trim the fat reflects a third of the American population who are overweight. But there is some encouraging news. Researchers have found that a protein produced by the mouse obesity gene, called Ob, when injected into mice, causes them to lose weight and maintain their weight loss. The Ob protein works by reducing appetite and speeding up metabolism. But as promising as these results are, they may not translate entirely to obese individuals since obesity is not as simple as a single flawed gene but many genes some of which don't respond to the Ob protein. Nevertheless, should the Ob work for humans, it could become the start of an effective way to burn more fat than dollars.

The New Skinny on Fat

A protein that makes mice shed pounds might ultimately help people, too

by Traci Watson

If Jenny Craig and Richard Simmons ever have job-related nightmares, the monster in their dreams must resemble the "ob mouse." The ob stands for obese, and, true to its name, the ob mouse is immense — up to three times as big as a regular mouse. Bred to be overweight, it is cursed with both a slow metabolism and a teenage boy's appetite.

But dietary salvation has now arrived for the ob mouse. In last week's issue of *Science*, three separate research teams announced the discovery of a slimming potion for the hapless mouse. The magic chemical is a protein called OB, and daily injections over several weeks were able to transform a naturally grotesque ob mouse into something approaching the mouse ideal.

The announcement caused a sensation in both scientific and financial circles and triggered a flurry of headlines about miracle diet pills. "These are very exciting results." Says Claude Bouchard, an obesity researcher at Laval University in Quebec. "They're going to keep us talking for a while."

But most researchers cautioned that it's too early to know for sure if OB could really contribute to a human obesity treatment. They caution that the protein will have to undergo years of testing before humans can take it — and it cannot be made into something as simple and as easy to take as a pill. Even then, OB may have side effects, and some scientists say it's never likely to provide a post-holiday or pre-college-reunion fat cure.

There are several things that make obesity researchers hopeful they are on the right track with OB, however. For one, the protein appears to work by both diminishing appetite and speeding up metabolism. Scientists speculate that the protein's natural role in the body is to tell the brain when the body's fat stores are growing: the brain, in turn, commands the body to lower food intake.

The ob mouse has been lolling around laboratories since the 1950s, when it first appeared by chance in mouse breeding colonies. But not until the 1960s and `70s did a series of grotesque experiments show that a biological compound could cause an ob mouse to lose weight. Douglas Coleman, a scientist at the Jackson Laboratory, the world-famous mouse-breeding facility in Bar Harbor, Maine, surgically coupled an ob mouse and a normal mouse to create a rodent version of Siamese twins. The fat half of the pair soon became skinny. Coleman realized that some kind of bloodborne factor must regulate weight. But the scientific techniques of Coleman's day were too blunt to find the flab-controlling substance.

In 1987, with molecular biology techniques much improved, scientists from the Howard Hughes Medical Institute at Rockefeller University began a concerted effort to find the gene responsible for the ob mouse's plight. It took them nearly eight years, but they finally tracked it down, and announced their discovery last December. The chunky mice, it turned out, all had a mutation in one gene, also designated ob. In these mutant mice, the ob gene malfunctions and fails to produce the OB protein. These mice grow large in part because their brains never get the message that the body already has lots of fat. As a result, the brain never tells the mouse to stop eating.

The clincher. Having found the ob gene and the OB protein it produces in normal mice, the Hughes-Rockefeller scientists and teams from the drug company Hoffmann-La Roche and the biotech firm Amgen performed the obvious experiment to clinch the connection: They gave injections of OB protein to the naturally fat ob mice. Then they sat back and watched the ounces melt away. Not only were the mice eating less but their metabolism was also speeding up. They lost weight; their fat stores shrank; their blood sugar, which bordered on diabetic levels, fell.

Why all this happened is still not precisely understood. The OB protein appears to have many roles in the body, and scientists aren't even sure which organs it targets. One effect does seem sure: OB shots finally tell the ob mouse's brain to shut down feeding. "This animal is fat because he thinks he's starving," says Jeffrey Friedman, leader of the Hughes-Rockefeller team. "When you give him the protein, the animal gets thinner because he thinks he's getting fatter."

Even more promising to those hoping for a practical fat therapy was OB's effect on so-called diet-induced obese mice. These are naturally svelte mice bred to have a fat tooth — and to gain weight quickly on a high-fat diet. So the Hoffmann-La Roche team fed their mice what team leader L. Arthur Campfield compares to "chocolate-chip cookie dough without the chocolate chips." Faced at every meal with the mouse equivalent of cheesecake, the mice overate and grew quite large. But when given shots of OB protein, the mice slimmed down in a hurry.

Also cheering was the finding that even lean mice unfattened by genes or diet lost weight on the OB diet plan. And at least some preliminary hints that what works for mice might work for men were further cause for optimism. The Hughes-Rockefeller team had already found last December that humans carry the ob gene. When the team gave *human* OB protein to ob mice, the mice still lost weight. So the human form of OB must work in a similar way to the mouse form.

A mutation in the human ob gene, researchers speculate, may be at the root of at least some cases of human obesity. These are the people that most scientists agree could almost certainly be helped by treatment with OB. But these would probably be people who are morbidly obese — weighing 300 to 400 pounds, a condition that is extremely rare.

People who are mildly obese, on the other hand, are hardly rare at all: A recent government survey found that one third of Americans are overweight. The reasons for that statistic are both environmental and genetic. Americans live in the land of milkshakes and honey-dipped doughnuts — and physical inactivity. Our genes prompt us to gain weight on this regimen. Thirty to 40 percent of obesity is due to genetics, scientists estimate. Indeed, a recent editorial in the *New England Journal of Medicine* dismissed as a "folk belief" the idea that overeating alone causes obesity.

Roadblocks. Whether OB can help people whose obesity is due to both genetics and lifestyle is up in the air. On the one hand, the weight loss that the diet-induced obese mice experienced on OB was promising. But fatness takes a thousand forms; there are a half-dozen known genes causing obesity in mice. And at least some of these forms of genetic obesity would definitely not be helped by the OB treatment.

The db (diabetes) mouse is a case in point. These mice are every bit as fat as ob mice, but when injected with the OB protein, the db mice stayed fat. Researchers suspect that these mice are OB "insensitive" — that is, their cells lack the ability to notice whether the OB protein is present. So even though the mice make their own OB, their bodies do not heed its effects, and extra protein doesn't help. Some overweight humans may very well have a similar problem, rendering them impervious to the powers of surplus OB.

Many other hurdles loom. At present, OB can be taken only by daily injection. Amgen, which is pursuing OB as a drug, hopes to come up with another way patients could take the drug — perhaps an implant or pump. An oral form is not possible, however, since the OB protein molecule is broken down by the digestive process before it can be absorbed through the gut. Worse, OB's effects last only as long as the drug is taken religiously: One research team took their mice off OB protein for only two days — and watched the mice's weight shoot up.

Side effects could also prevent OB from entering the mass market. The protein seems likely to be a hormone, and hormones, such as insulin, often have wide-ranging effects on the body. If this is the case for OB, long-term use of the drug could spring unpleasant surprises. Mice given too much OB protein waste away and die. Tests for side effects in humans should start in 1996, followed by tests of OB's effectiveness. If all goes well, OB could enter the market in five to seven years.

Luckily, if OB doesn't pan out, scientists have lots of other ideas, many of which involve pills. It may be possible, for example, to find a chemical that boosts the body's own production of OB. Researchers are also investigating ways to turn up the body's sensitivity to OB, which would benefit db mice and any humans with a similar genetic defect. But these drugs would require at least an extra five years to develop.

Other therapeutic avenues may be opened as more obesity genes are discovered. The latest fat gene was announced last week by a separate group of researchers. Officials at Millennium Pharmaceuticals of Cambridge, Mass., said their scientists had found the mouse and human versions of a gene dubbed "tub." Tubby mice, as they're known, never get quite as Rubenesque as ob mice. And while the ob mouse is fat from Day One, the tubby mouse fattens slowly after reaching puberty. That's how most people gain weight, too, so scientists are anxious to find out more about how tub acts.

Doctors worry, of course, about misuse of fat-melting miracle proteins. But even the worriers think the benefits outweigh the hazards because overweight people who slim down reduce their risk of diabetes and heart disease. And the ob mice saw another benefit, too, becoming livelier and more active — yet without losing their sweet, gentle natures. No human dieter could ask for more. ❑

****

Questions:

1. How does the Ob protein work in the body?
2. Why must the Ob protein be injected?
3. Will the Ob protein work for all forms of genetic obesity?

Answers are at the end of the book.

Watching your weight is important for good health and so is your diet. Studies have been done to demonstrate a connection between lifestyle and risks of breast cancer. Particular foods which offer some protection are olive oil or oils rich in unsaturated fatty acids, fruits, vegetables, fiber and soybeans. Although this information is derived from observational studies, it does help to know we can make enlightened choices about the foods we eat and possibly help to prevent breast cancer.

Diet and Breast Cancer

Harvard Women's Health Watch

When it comes to breast cancer risk, the news may be getting better. No longer does it seem as though the only factors that influence risk are things we can do little about — family history, having children late or not at all, early menarche, and late menopause; lately, observational studies have begun to demonstrate that diet also may play an important role.

One of the most recent reports came from Greece, where breast-cancer incidence is lower than in the United States. A team of researchers from the University of Athens and Harvard School of Public Health found that women who consumed olive oil at more than one meal a day had a significantly lower risk of breast cancer than those who used olive oil less frequently. Fruits and vegetables also appeared to have a protective effect.

The study, which appeared in the January 18, 1995, issue of the *Journal of the National Cancer Institute*, with more than 2300 participants, was one of the largest to zero in on specific categories of foods. Women who had breast cancer were matched with women who had similar characteristics but had not developed the disease, and all were asked to report how often they ate several types of foods, including margarine, olive oil, and animal fats.

As was the case in many other studies looking for links between lifestyle and cancer, the information depended upon the women's ability to accurately report the foods they ate. Because memory is fallible, such observational studies are considered to be less definitive than controlled trials in which two groups of women with similar characteristics are placed on different diets and the number of new breast cancers in each group is compared. Nonetheless, observational studies are valuable because they not only focus the direction of future clinical trials but also serve as interim guidance for those of us trying to do what we can to prevent breast cancer.

To date, observational studies have indicated certain connections between breast cancer and the following foods:

• *Dietary fat.* The evidence is conflicting. Although some studies show breast cancer risk increasing with fat intake, the overall breast cancer rate is significantly lower among women in the Arctic North and the Mediterranean, where fat accounts for more than 40% of calories, than among American women, who consume, on average, 36% of calories as fat.

In both of the latter populations, fat is provided primarily by fish oils or olive oil, both of which are rich in unsaturated fatty acids. In contrast, dietary fat for most Americans is derived principally from saturated fatty acids in meat and dairy products and from hydrogenated polyunsaturated oils containing *trans*-fatty acids, which are found in margarine and packaged snacks and pastries. Because there is no indication that a low-fat diet increases breast cancer risk, a prudent course would be to restrict fat to 30% of calories, as recommended by the National Research Council, by reducing saturated fats and *trans*-fatty acids.

• *Dietary fiber.* A high fiber intake is consistently associated with a lowered risk of breast cancer. Insoluble fibers like bran, derived principally from unprocessed grains, seem to provide the greatest protective effect; soluble fibers from fruits and vegetables also seem to confer protection.

Fiber intake appears to reduce levels of circulating

estrogens, which may promote the growth of breast tumors, perhaps by curtailing the production of liver enzymes that foster estrogen absorption and by binding to the estrogens in the intestines.

• *Soybeans,* which contain cancer-preventing compounds called isoflavonoids, are also a source of phytoestrogen, which is chemically similar to estrogen. By replacing estrogen on certain tissue receptors, phytoestrogens may block its cancer-promoting effects. ❑

****

Questions:

1. What level of fat is acceptable in the diet according to the National Research Council?

2. What health benefit does fiber provide?

3. What cancer-preventing compound is found in soybeans?

Answers are at the end of the book.

An aspirin a day keeps the doctor away? Aspirin, the world's most widely used drug, is under intense investigation for its uses in disease treatment and prevention. In laboratories and clinics, a new understanding of how aspirin achieves some of its effects by inhibiting the production prostaglandin help to explain why taking aspirin can reduce fever and ease minor aches that occur with inflammation. Scientists are most interested in the ways aspirin can be used to prevent cardiovascular disease and are designing the experiments to test the very idea. Until the results are in, health officials are warning consumers to practice caution and consult a doctor before taking aspirin for new and long-term uses.

ASPIRIN - A New Look at an Old Drug

Americans consume an estimated 80 billion aspirin tablets a year.

by Ken Flieger

In purses and backpacks, in briefcases and medicine chests the world over, millions of people keep close at hand a drug that has both a long past and a fascinating future. Its past reaches at least to the fifth century B.C., when Hippocrates used a bitter powder obtained from willow bark to ease aches and pains and reduce fever. Its future is being shaped today in laboratories and clinics where scientists are exploring some intriguing new uses for an interesting old drug.

The substance in willow bark that made ancient Greeks feel better, salicin, is the pharmacological ancestor of a family of drugs called salicylates, the best known of which is the world's most widely used drug—aspirin.

Americans consume an estimated 80 billion aspirin tablets a year. The *Physicians' Desk Reference* lists more than 50 over-the-counter drugs in which aspirin is the principal active ingredient. Yet, despite aspirin's having been in routine use for nearly a century, both scientific journals and the popular media are full of reports and speculation about new uses for this old remedy. The National Library of Medicine's main computerized catalog includes more than 2,700 scientific articles about aspirin. And those are only the English language publications that have appeared in the last five years.

Yet aspirin's beginnings were rather unspectacular. Nearly 100 years ago, a German industrial chemist, Felix Hoffmann, set about to find a drug to ease his father's arthritis without causing the severe stomach irritation associated with sodium salicylate, the standard anti-arthritis drug of the time. In the forms then available, the large doses of salicylates used to treat arthritis — 6 to 8 grams a day — commonly irritated the stomach lining, and many patients, like Hoffmann's father, simply could not tolerate them.

Figuring that acidity made salicylates hard on the stomach, Hoffmann started looking for a less acidic formulation. His search led him to synthesize acetylsalicylic acid (ASA), a compound that appeared to share the therapeutic properties of other salicylates and might cause less stomach irritation. ASA reduced fever, relieved moderate pain, and, at substantially higher doses, alleviated rheumatic and arthritic conditions. Hoffmann was confidant that ASA would prove more effective than salicylates then in use.

His superiors, however, did not share his enthusiasm. They doubted that ASA would ever become a valuable, commercially successful drug because at large doses salicylates commonly produced shortness of breath and an alarmingly rapid heart rate. It was taken for granted — incorrectly as it turns out — that ASA would weaken the heart and that physicians would be reluctant to prescribe it in preference to sodium salicylate, a drug they at least knew. Hoffmann's employer, Friedrich Bayer & Company, gave ASA the now-familiar name aspirin, but in 1897 Bayer didn't think aspirin had much of a future. It could not have foreseen that almost a century after its development aspirin would be the focus of extensive laboratory research and some of the largest

Reprinted from *FDA Consumer*, January-February 1994.

clinical trials ever carried out in conditions ranging from cardiovascular disease and cancer to migraine headache and high blood pressure in pregnancy.

How Does It Work?

The mushrooming interest in aspirin has come about largely because of fairly recent advances in understanding how it works. What is it about this drug that, at small doses, interferes with blood clotting, at somewhat higher doses reduces fever and eases minor aches and pains, and at comparatively large doses combats pain and inflammation in rheumatoid arthritis and several other related diseases?

The answer is not yet fully known, but most authorities agree that aspirin achieves some of its effects by inhibiting the production of prostaglandins. Prostaglandins are hormone-like substances that influence the elasticity of blood vessels, control uterine contractions, direct the functioning of blood platelets that help stop bleeding, and regulate numerous other activities in the body.

In the 1970s, a British pharmacologist, John Vane, Ph.D., noted that many forms of tissue injury were followed by the release of prostaglandins. In laboratory studies, he found that two groups of prostaglandins caused redness and fever, common signs of inflammation. Vane and his co-workers also showed that, by blocking the synthesis of prostaglandins, aspirin prevented blood platelets from aggregating, one of the initial steps in the formation of blood clots.

This explanation of how aspirin and other nonsteroidal anti-inflammatory drugs (NSAIDs) produce their intriguing array of effects prompted laboratory and clinical scientists to form and test new ideas about aspirin's possible value in treating or preventing conditions in which prostaglandins play a role. Interest quickly focused on learning whether aspirin might prevent the blood clots responsible for heart attacks.

A heart attack or myocardial infarction (MI) results from the blockage of blood flow not *through* the heart, but *to* heart muscle. Without an adequate blood supply, the affected area of muscle dies and the heart's pumping action is either impaired or stopped altogether.

The most common sequence of events leading to an MI begins with the gradual build-up of plaque (atherosclerosis) in the coronary arteries. Circulation through these narrowed arteries is restricted, often causing the chest pain known as angina pectoris.

An acute heart attack is believed to happen when a tear in plaque inside a narrowed coronary artery causes platelets to aggregate, forming a clot that blocks the flow of blood. About 1,250,000 persons suffer heart attacks each year in the United States, and some 500,000 of them die. Those who survive a first heart attack are at greatly increased risk of having another.

Could Aspirin Help?

To learn whether aspirin could be helpful in preventing or treating cardiovascular disease, scientists have carried out numerous large randomized controlled clinical trials. In these studies, similar groups of hundreds or thousands of people are randomly assigned to receive either aspirin or a placebo, an inactive, look-alike tablet. The participants — and in double-blink trials the investigators, as well — do not know who is taking aspirin and who is swallowing a placebo.

Over the last two decades, aspirin studies have been conducted in three kinds of individuals: persons with a history of coronary artery or cerebral vascular disease, patients in the immediate, acute phases of a heart attack, and healthy men with no indication of current or previous cardiovascular illness.

The results of studies of people with a history of coronary artery disease and those in the immediate phases of a heart attack have proven to be of tremendous importance in the prevention and treatment of cardiovascular disease. The studies showed that aspirin substantially reduces the risk of death and/or non-fatal heart attacks in patients with a previous MI or unstable angina pectoris, which often occurs before a heart attack.

On the basis of such studies, these uses for aspirin (unstable angina, acute MI, and survivors of an MI) are described in the professional labeling of aspirin products, information provided to physicians and other health professionals. Aspirin labeling intended for the general public does not discuss its use in arthritis or cardiovascular disease because treatment of these serious conditions — even with a common over-the-counter drug — has to be medically supervised. The consumer labeling contains a general warning about excessive or inappropriate use of aspirin, and specifically warns against using aspirin to treat children and teenagers who have chickenpox or the flu because of the risk of Reye syndrome, a rare but sometimes fatal condition.

Aspirin for Healthy People?

Once aspirin's benefits for patients with cardiovascular disease were established, scientists sought to learn whether regular aspirin use would prevent a first heart attack in healthy individuals. The findings regarding that critical question have thus far been equivocal. The major American study designed to find out if aspirin can prevent cardiovascular deaths in healthy individuals was a randomized, placebo-controlled trial involving just over 22,000 male physicians between 40 and 84 with no prior history of heart disease. Half took one 325-milligram aspirin tablet every other day, and half took a placebo.

The trial was halted early, after about four-and-a-half years, and the findings quickly made public in 1988 when investigators found that the group taking aspirin had a substantial reduction in the rate of fatal and non-fatal heart attacks compared with the placebo group. There was, however, no significant difference between the aspirin and placebo groups in number of strokes (aspirin-treated patients did slightly worse) or in overall deaths from cardiovascular disease.

A similar study in British male physicians with no previous heart disease found no significant effect nor even a favorable trend for aspirin on cardiovascular disease rates. The British study of 5,100 physicians, while considerably smaller than the American study, reported three-quarters as many vascular "events." FDA scientists believe the results of the two studies are inconsistent.

The U.S. Preventive Services Task Force, a panel of medical-scientific authorities in health promotion and disease prevention, is one of many groups looking at new information on the role of aspirin in cardiovascular disease. In its *Guide to Clinical Preventive Services*, issued in 1989, the task force recommended that low-dose therapy "should be considered for men aged 40 and over who are at significantly increased risk for myocardial infarction and who lack contraindications" to aspirin use. A revised *Guide*, scheduled for publication in the fall of 1994, is expected to include a slightly revised recommendation concerning aspirin and cardiovascular disease but no major change in advice to physicians about aspirin's possible role in preventing heart attacks.

Better understanding of aspirin's myriad effects in the body has led to clinical trials and other studies to assess a variety of possible uses: preventing the severity of migraine headaches, improving circulation to the gums thereby arresting periodontal disease, preventing certain types of cataracts, lowering the risk of recurrence of colorectal cancer, and controlling the dangerously high blood pressure (called preeclampsia) that occurs in 5 to 15 percent of pregnancies.

None of these uses for aspirin has been shown conclusively to be safe and effective, and there is concern that people may be misusing aspirin on the basis of unproven notions about its effectiveness. Last October, FDA proposed a new labeling statement for aspirin products advising consumers to consult a doctor before taking aspirin for new and long-term uses. The proposed statement would read, "IMPORTANT: See your doctor before taking this product for your heart or for other new uses of aspirin because serious side effects could occur with self treatment."

The Other Side of the Coin

While examining new possibilities for aspirin in disease treatment and prevention, scientists do not lose sight of the fact that even at low doses aspirin is not harmless. A small subset of the population is hypersensitive to aspirin and cannot tolerate even small amounts of the drug. Gastrointestinal distress — nausea, heartburn, pain — is a well-recognized adverse effect and is related to dosage. Persons being treated for rheumatoid arthritis who take large daily doses of aspirin are especially likely to experience gastrointestinal side effects.

Aspirin's antiplatelet activity apparently accounts for hemorrhagic strokes, caused by bleeding into the brain, in a small but significant percentage of persons who use the drug regularly. For the great majority of occasional aspirin users, internal bleeding is not a problem. But aspirin may be unsuitable for people with uncontrolled high blood pressure, liver or kidney disease, peptic ulcer, or other conditions that might increase the risk of cerebral hemorrhage or other internal bleeding.

New understanding of how aspirin works and what it can do leaves no doubt that the drug has a far broader range of uses than Felix Hoffmann and his colleagues imagined. The jury is still out, however, on a number of key questions about the best and safest ways to use aspirin. And until some critical verdicts are handed down, consumers are well-advised to regard aspirin with appropriate caution. ❑

****

Questions:

1. What is aspirin?

2. How does aspirin appear to work?

3. What are aspirin's possible roles in treating and preventing disease?

Answers are at the end of the book.

Part Seven:

Controversies and Ethical Issues

Human embryos are in demand by infertile couples who can't conceive a child and for their potential use as a method of genetic diagnosis. But the idea of "cloning" a human embryo seemed almost outrageous even to the scientific community when it was announced by two researchers in 1993. The experiment, reported by Jerry L. Hall and Robert J. Stillman involved the duplication of very young undifferentiated embryos and not the duplication of adult cells which have become specialized. Aside from the technical achievement, Hall and Stillman used the word "cloning" to describe their human embryo research which triggered an avalanche of horror since no scientist had dared to traverse this ethical line before. While the work is fundamental to reproductive medicine, it also stirs up the old ethical debate of whether human embryo cloning should be allowed at all.

Cloning Human Embryos

Exploring the science of a controversial experiment

by Kathy A. Fackelmann

On October 13, 1993, researchers Jerry L. Hall and Robert J. Stillman entered the annals of reproductive history with the announcement that they had "cloned" a human embryo.

Ethicists raised the specter of society straight out of Aldous Huxley's *Brave New World*, complete with scores of duplicate human babies. A Vatican theologian denounced the research effort as "perverse."

Meanwhile, infertile couples worried that a backlash could curtail funding for such experiments. The announcement put a "strange spotlight on infertility treatment," notes Diane Aronson, executive director of Resolve, an organization of infertile couples based in Somerville, Mass.

As for the scientific significance of the experiment, conflicting views abound. Many researchers working with human embryos consider the work laudable, noting that it may help infertile couples conceive a child. Others say "cloning" techniques may lead to an improved method of genetic diagnosis. Yet researchers working with animal embryos call the findings "ho-hum" and almost trivial in nature.

"Any graduate student could have done the experiment," says George E. Seidel Jr. of the Animal Reproduction Laboratory at Colorado State University in Fort Collins.

Almost all fertility specialists agree, however, that inaccuracies fueled the public response to the announcement. Most blame the news media and ethicists for conjuring up frightening scenarios of the future. At the same time, some fertility specialists admit that the research community muddied discussions of "cloning" by using technical terms in an imprecise manner.

This analysis attempts to elucidate some of the unexplored scientific issues raised by the first report of "cloning" a human embryo.

The word "clone" comes from a Greek word that means "twig" and suggests the practice of slicing off a piece of a plant and rooting it. Gardeners routinely use this practice to duplicate a favorite shrub.

One modern definition of "clone," as found in *Merriam-Webster's Collegiate Dictionary*, is "an individual grown from a single somatic cell of its parent and genetically identical to it." All body cells except those that give rise to sex cells are somatic.

Scientists have never taken such a nonreproductive cell from an adult human — or any other adult mammal — and fashioned an identical clone. Indeed, such a feat remains in the realm of fiction, at least for now.

Why? Adult cells are differentiated, or specialized, to perform a specific function. Differentiation is the developmental process by which unspecialized embryonic cells take on their mature role in the body. Once the process is complete, there's no turning back to an unspecialized state. An adult skin cell, for example, can't transform itself into an undifferentiated cell.

The experiment reported by Hall and Stillman, both at the

George Washington University Medical Center in Washington, D.C., fell far from Webster's definition of cloning. The two scientists duplicated very young embryonic cells, not adult cells, points out Seidel. Such cells have yet to specialize, he notes.

In addition, the research Hall and Stillman described last October at the joint meeting of the American Fertility Society and the Canadian Fertility and Andrology Society is quite different from the popular notion of cloning. For example, in the movie *Jurassic Park*, scientists used somatic cells that had been trapped in amber to create replicas of long-extinct dinosaurs.

Thus, when Hall and Stillman used the word "cloning" to describe their research, many people reacted as if the pair had fashioned an exact copy of an adult human, points out Howard W. Jones, Jr., honorary chairman and one of the founders of the Jones Institute for Reproductive Medicine at the Eastern Virginia Medical School in Norfolk. "*Jurassic Park* is simply science fiction," he says.

To understand what Hall and Stillman actually did accomplish, one must first consider the backdrop, including a long history of cloning by animal researchers.

In the 1940s and 1950s, embryologists took young embryos from rats and successfully separated each embryo into its few constituent cells. At that time, researchers knew that the egg starts to divide after fertilization, forming genetically identical cells, or blastomeres. A tough outer covering, the zona pellucida, protects the fragile blastomeres.

During the 1970s, researchers relied on the same technique, known as blastomere separation, to produce identical twin mouse pups.

Such work showed that each blastomere has the ability to develop into any type of cell. While an adult skin cell can never turn into a heart cell, a blastomere can become a skin cell, a heart cell, or any other cell in the body.

The next landmark occurred in 1979, when Steen Willadsen, then at the Institute of Animal Physiology in Cambridge, England, detailed a blastomere separation procedure for use on larger animals — in this case, sheep. He published his findings in the Jan. 25 NATURE.

Willadsen described removing very young embryos from ewes. With an extremely fine needle, he poked a hole in the zona pellucida, then sucked the blastomeres out. To provide some protection, he coated each "naked" blastomere with a gelatinous material called agar.

When Willadsen transferred these agar-coated blastomeres to the womb of an ewe, they began to divide. Proof of the experiment's success came with the birth of several sets of identical twin lambs.

The NATURE paper represented a large leap forward, recalls Seidel. Soon after, other researchers employed blastomere separation to create twin lambs and calves.

Before long, the scientific terminology began to get messy. Although blastomere separation involves the isolation of embryonic cells, some scientists referred to the technique as a type of cloning, Seidel says.

The potential for confusion increased when scientists developed another reproductive technique — nuclear transplantation — also considered a type of cloning. This procedure, too, requires unspecialized embryonic cells and so far cannot be done with adult cells, Seidel says.

Nuclear transplantation works this way: Scientists obtain an embryo that has developed to the stage where it consists of 32 blastomeres. They separate these blastomeres, which contain identical genetic material. They then use an electric current or some other method to coax each blastomere to fuse with an egg cell whose nucleus has been removed.

In theory, nuclear transplantation could yield hundreds of identical high-volume dairy cows or other domestic animals with blue-ribbon qualities, Seidel says. In reality, the technique has fallen far short of that goal, he adds.

Hall and Stillman started their experiment with 17 very young human embryos slated for discard at an infertility clinic. All had started dividing and consisted of two to eight blastomeres. Using an enzyme called pronase, the duo dissolved the zona pellucida and separated the blastomeres. Next, they coated each blastomere with a synthetic shell made of a material derived from seaweed. They allowed those cells to develop in a laboratory dish, noting that some of the blastomeres divided a few times and then died.

Blastomeres from the two-celled embryos did best of all. Some made it to 32-cell divisions, a stage at which they could be transferred to the womb, Hall says.

Stillman says the method they used goes by a variety of names, including "twig cloning," "embryo twinning," or simply "cloning." Yet Seidel points out that their technique is probably most accurately described as blastomere separation.

Many scientists now regret this widespread lack of rigor in describing such complex methods. The public confusion over Hall and Stillton's research graphically illustrates the importance of using terms that describe exactly what was done, comments Mary C. Martin of the University of California, San Francisco (UCSF). "I think we should be very strict in our terminology," adds Robert G. Edwards, the *in vitro* fertilization pioneer whose work, along with that of Patrick Steptoe, led to the world's first test-tube baby in 1978.

In one sense, the Hall-Stillman experiment was designed to fail. The researchers used polyspermic embryos, which result when more than one sperm penetrates an egg. Such abnormal embryos have too much genetic information and cannot survive. Hall and Stillman turned to these flawed embryos, which had been slated for routine disposal, because testing normal human embryos in such a preliminary experiment would have been unethical, they said.

In another sense, the pilot study proved a success. It suggested that if one applied the same methods to normal human embryos, one could obtain viable blastomeres that would develop in the uterus, Seidel says.

Furthermore, there's no reason why such methods wouldn't work if carried to their logical conclusion: the transfer of such artificially coated blastomeres to a woman's womb and the birth of identical twins or triplets, he adds.

A review of the animal research shows that scientists have had the technical expertise to clone a human embryo for years. Why didn't they forge ahead? Most scientists cite ethical reasons for the *de facto* moratorium on "cloning" human embryos.

Why did Hall and Stillman break through that barrier? At a news conference in Washington, D.C., last October, Hall said they did the experiment to spur an ethical debate on the value of cloning human embryos.

The text of their scientific abstract, however, doesn't mention that as a goal. It states, "This technique could be useful to patients who have difficulty producing sufficient numbers of embryos for transfer." In fact, Hall now says that he considered the ability to create identical twins a scientific challenge — one that could provide substantial benefits to infertile couples.

Women who produce one egg have a 10 percent chance of a successful pregnancy, Stillman explains. If researchers could multiply a single egg, the pregnancy rate would increase dramatically, he says.

Most fertility scientists see nothing wrong with that application of the research, noting that nature produces identical twins in much the same way: The two-celled embryo divides and eventually develops into two identical babies.

Indeed, says Gary D. Hodgen, president of the Jones Institute, if this reproductive technique can help infertile couples conceive a child, then this research would be "completely justified."

Researchers judging the scientific abstracts submitted for presentation at the joint fertility meeting had been impressed enough by Hall and Stillman's work to award it the top prize. The George Washington team's abstract, ranked blindly by two separate peer review panels, beat 90 others for the honor.

Scientists who attended the meeting, which was held in Montreal, reacted favorably to the abstract. "It was an important study," recalls UCSFs Martin. "It's a nice piece of work," concurs Edwards, who is now a professor emeritus at the University of Cambridge in England.

Some scientists say that one of the chief applications of the new method got lost in the media uproar. Lucinda L. Veeck, also at the Jones Institute, says that Hall and Stillman's technique would boost the efficiency of a new form of genetic diagnosis, one that can tell prospective parents whether a tiny embryo has inherited a serious disease, such as Tay-Sachs, cystic fibrosis, hemophilia, or muscular dystrophy.

With preimplantation diagnosis, couples with a family history of a serious genetic disease can find out an embryo's risk before it is transferred to the womb. The technique involves the now standard procedure of uniting a human egg and sperm in a petri dish. Once the fertilized egg begins to divide, researchers punch a hole in the zona pellucida and suck out a single blastomere. They then analyze the blastomere's DNA, searching for signs of an inherited disorder.

But researchers can't always get enough DNA from a single blastomere to make an accurate diagnosis, Veeck says. If they multiplied that single blastomere using Hall and Stillman's cloning methods, researchers would have a bigger pool of DNA — and a better chance of predicting the future, she says.

The artificial zona pellucida was another scientifically notable aspect of the controversial "cloning" research, Martin says. She points out that Hall had won the top prize at the American Fertility Society meeting in 1991 for developing and testing the jelly-like coating on mouse embryos.

Without a protective "shell," fragile human blastomeres would die, never developing into an embryo, Hall says. Thus, the team's successful use of the seaweed-derived zona paved the way for more sophisticated experiments — such as the transfer of coated embryos to an infertile woman's womb, he says.

Hall envisions another use for the artificial zona: He believes the material could be used to repair damage to an egg's protective coating during test-tube fertilization.

Not everyone views that aspect of the report with enthusiasm. *In vitro* fertilization researcher Jacques Cohen says the development of a synthetic shell is just another "bell and whistle," not something really necessary to hold an early embryo together.

"You don't need an artificial zona," says Cohen of Cornell University Medical Center in New York City. Zona-free blastomeres taken from mouse embryos will continue to develop if they're left untouched in a laboratory dish, he says.

Cohen speculates that Hall and Stillman received the top prize at the fertility meeting not because of their abstract's technical merits, but because the reviewers were unduly impressed with its fancy terminology, such as the use of the word "cloning."

"There were a lot of gimmicky tricks in this paper to make it look sexy," he says. In addition, he believes, the reviews failed to take into account the vast animal research that had gone before. Indeed, the abstract seems little more than an updated version of Willadsen's work with sheep, an experiment reported 14 years earlier.

It comes as no surprise, then, that animal researchers have greeted this "advance" with muted enthusiasm.

The procedure was not as advanced as nuclear transplantation experiments being done routinely with animal embryos, says Willadsen, who is now a researcher in Calgary, Alberta. And Neal First, a reproductive biologist at the University of Wisconsin-Madison, calls the hullabaloo over the human "cloning" experiment "a lot of fuss over nothing."

Such comments could be construed as sour grapes, Willadsen acknowledges, adding that there will always be competition between animal researchers and those working in the human arena. Research with animal embryos can far outstrip the technical achievements of experiments with human embryos, he adds, largely because animal research is unfettered by the same ethical constraints as research on humans.

The news that Hall and Stillman had cloned human embryos raised a welter of complex scientific and ethical issues. Some scientists argue that the cloning report represents a significant advance that promises new hope for infertile couples. Others say the report simply rehashed the work that had already been done with animal embryos.

No one would deny, however, that one key aspect of the report was simply Hall and Stillman's use of human embryos. No scientist had dared cross that ethical boundary before, even though the technology had existed for years.

"The human embryo is considered the sacred sanctum," Willadsen says, adding that, despite the controversy, such work should go forward.

Veeck recalls the early days of *in vitro* fertilization when "busloads of angry protesters" opposed the practice of uniting human sperm and egg in a test tube. Yet work with *in vitro* fertilization went forward, helping thousands of couples deliver healthy children, she notes.

Many scientists worry that mounting ethical concerns triggered by the rapid-fire advances in reproductive medicine could bring such research to a halt. Indeed, Hall and Stillman's report was soon followed by the news that an Italian scientist news that an Italian scientist had used *in vitro* techniques to help a 62-year-old woman become pregnant, a move that prompted French government officials to propose banning the procedure.

Edwards, who has had plenty of experience with such debate, argues that the very process of wading through a thicket of ethical questions will prove beneficial to society, forcing it to cope with almost undreamed of technical advances.

"You're always going to get these arguments, because early human life is a very precious thing," he says.❑

****

Questions:

1. Why did Hall and Stillman's experiment cause such controversy?

2. What are scientists able to clone?

3. Why did Hall and Stillman perform the experiment?

Answers are at the end of the book.

What is the future of the US fetal tissue bank? With tissue transplantation on the rise, it is becoming increasingly necessary to re-establish a tissue bank which would collect, process, and distribute this tissue throughout the country. Before this can occur, many ethical, scientific and public policy issues must be worked out to insure its efficacy and stability. For example, a tissue bank of this nature has raised ethical concerns about adding to the system the use of fetal tissue from elective abortions. Also, guidelines will have to be set to ensure the safety and suitability of the tissue to prevent the spread of potentially fatal diseases such as HIV to the recipient. Once these criteria are met, it will provide a standard by which to furnish the fetal tissue needed for transplantation research in the United States.

The Future of the Fetal Tissue Bank

by Cynthia B. Cohen and Albert R. Jonsen (for the National Advisory Board on Ethics Reproduction)

The fetal tissue bank established by the Presidential Executive Order of May 1992 was canceled in a little noticed clause in the 1993 National Institutes of Health (NIH) Revitalization law (1). Should the fetal tissue bank be allowed to slip into obscurity? Or should it be revived? If revived, what goals should it pursue?

The fetal tissue bank was planned to develop in two phases. Feasibility studies would first be conducted to evaluate whether sufficient uninfected, viable, and cytogenetically normal fetal tissue could be retrieved from ectopic pregnancies and spontaneous abortions to meet transplantation research needs. If an adequate amount of material could be made available, a fetal tissue bank would then be established to collect, process, and distribute this tissue through regional centers around the country. The bank would be composed of a coordinated network of retrieval centers, rather than one central collection agency. Five investigators awarded 2-year peer-reviewed grants by the National Institute of Child Health and Human Development (NICHD) carried out the first phase of the project (2). However, a decision was made in mid-1993 by the Department of Health and Human Services (HHS) to drop plans for the fetal tissue bank. This decision was reinforced by passage of the NIH Revitalization law.

The main reason given for abandoning the fetal tissue bank was that such a bank is needed only when large quantities of fetal tissue are being used (3). As fetal tissue transplantation is still in the experimental stages and requires only small amounts of tissue, it was argued, there is no need to set up a fetal tissue bank at this time. Further, some transplant investigators indicated that they preferred to dissect and process fetal tissue themselves, rather than use tissue retrieved and tested by others (3).

No published data are available to indicate the optimal quantity of fetal tissue required to meet transplant research needs each year. Although various estimates form the basis of clinical transplantation protocols, such estimates are often determined by the amount of tissue that is available (4). One American investigator who uses fetal brain tissue implants for patients with Parkinson's disease indicated that his work required approximately 200 fetuses a year (5). Transplants of fetal pancreas tissue in diabetes research have used tissue from 6 to 18 fetuses of 16 to 20 weeks gestational age per recipient (6). If only 10 patients with diabetes per year throughout the United States were to receive transplants of fetal tissue, this research could require as many as 180 fetuses. When the research requirements of other fetal tissue transplant investigators are taken into account, many thousands of sources of fetal tissue could be needed per year. This estimate is confirmed by statements of transplant investigators who objected to the establishment of the fetal tissue bank because it could not supply the several thousand fetuses needed every year for their research (7).

The primary reason for

Albert R. Jonsen, Ph.D., Chair, National Advisory Board on Ethics in Reproduction (NABER) and Cynthia B. Cohen, Ph.D., J.D., Senior Research Fellow at Kennedy Institute of Ethics, Georgetown University, Washington, D.C.

establishing a fetal tissue bank, however, is not to facilitate the distribution of large amounts of fetal tissue, but to satisfy ethical concerns. A fetal tissue bank provides a barrier between those who undergo and carry out abortions and those who receive and perform fetal tissue transplants. The presence of this wall of separation lessens the possibility that women who would not otherwise have an abortion will be influenced to do so. It also decreases the chance that conflicts of interest or collusion will occur among those involved in transplant investigations and those carrying out abortions. A fetal tissue bank, in addition, serves as a check on the uniformity, quality, and safety of fetal tissue provided to transplant recipients. Such a bank can develop model standards for quality control and equitable distribution of this tissue.

The Polkinghome Report, a document developed by a government-appointed committee in the United Kingdom in 1989, recommended the separation of those involved in fetal tissue donation from those who use such tissue. It maintained that this separation is best achieved through the establishment of an intermediary organization (8). A fetal tissue bank was therefore established in the United Kingdom, even though the quantity of fetal tissue being used for transplantation research in that country was limited and research was in its early stages (9). The British fetal tissue bank undertook to collect and process fetal tissue rapidly, using uniform methods developed in cooperation with investigators (4, 9). For similar reasons of ethics and to ensure an adequate supply of tissue of good quality, we believe that the American fetal tissue bank should not be abandoned.

Ethical Issues

There are good reasons for using fetal tissue from elective abortions in a U.S. fetal tissue bank. This tissue is more plentiful than that from spontaneous abortions and ectopic pregnancies and is less likely to be infected and genetically abnormal. Special ethical concerns, however, have been raised about the use of fetal tissue from elective abortions. The 1988 Panel on Human Fetal Tissue Transplantation Research concluded that it is acceptable public policy to use fetal tissue obtained from elective abortions for medical purposes, provided that certain ethical guidelines are followed (10). The recommendations of the Fetal Tissue Panel have the effect of separating the decision about abortion from the decision to donate fetal tissue, thereby alleviating concern voiced by some about direct complicity in abortion by those who use this tissue. They remove incentives for inducing abortion by prohibiting the sale of fetal tissue and directed donation of this tissue to a designated individual. They indicate that the timing and method of abortion should not be influenced by the potential use of fetal tissue and that the consent of the woman who donates fetal tissue must be obtained before the tissue can be used. Although the NIH Revitalization Act specifically dropped the fetal tissue bank, it included certain requirements about fetal tissue use that reflect the recommendations of the 1988 Fetal Tissue Panel.

These recommendations were developed 5 years ago in a rapidly advancing field on the basis of information and capabilities existing at that time. Their purview was limited by the restricted set of questions given the Fetal Tissue Panel by the Secretary of HHS. In our opinion, another panel should be convened to review recent developments in the field and to develop expanded guidelines for fetal tissue use.

A new fetal tissue panel would need to readdress the scope of the principle of separation. Fetal tissue currently used in transplantation, which is derived from the brain, pancreas, and liver, needs to be removed in a way that renders it identifiable if it is to be usable. A new panel should consider whether changes in the timing and method of abortion that are not significant and that would retain the integrity of the tissue without placing the woman at greater risk would constitute an unacceptable breach of the wall of separation. A new panel should also reconsider the conclusion of the 1988 panel that the consent of the woman to donate fetal tissue is sufficient for the use of that tissue "unless the father objects." There is no legal obligation to obtain the permission of the male partner for a woman's abortion. To open the door to requiring his permission for the donation of fetal tissue after abortion has taken place might place the donor at risk. This recommendation of the panel was changed in the provisions of the NIH Revitalization Act of 1993 pertinent to research on transplantation of fetal tissue. It is important that a new panel elaborate the reasons why this change is ethically sound. Another issue that needs to be addressed is the availability of counseling for women who donate fetal tissue. Both fetal tissue and the women who donate it must be tested to evaluate the risk of infection to the recipient. The results of this testing could have serious implications for the health of the donor. Therefore, appropriate forms of counseling should be made available for donors.

Although a fetal tissue bank in the United States should include tissue from elective abortions, it should not be restricted to this tissue alone. Some investigators and patients, for reasons of conscience, will not participate in research using tissue from this source. In an open, pluralistic society such as ours, their perspective should be accommodated and, if feasible, tissue from spontaneous abortions and ectopic pregnancies should be included in a fetal tissue bank. Potential recipients of fetal tissue implants who have a preference for tissue from sources other than elective abortions should be informed of the advantages and disadvantages of using tissue from all sources and allowed to indicate their preference. If they request tissue from spontaneous abortions or ectopic pregnancies, they should be provided with it if it is available and has been thoroughly screened.

Scientific and Public Policy Issues

Concerns have been raised about the possibility of a major increase in unregulated and undocumented agreements for the acquisition of fetal tissue in the aftermath of the dissolution of the moratorium on the use of fetal tissue from elective abortions (11, 12). A large-scale commercial fetal tissue industry could emerge in this country. Yet no guidelines are in place for obtaining, testing, processing, freezing, and storing fetal tissue in the United States. No organized system is being planned to distribute this tissue on a nonprofit and equitable basis. In contrast, the distribution of solid organs for transplantation is coordinated through a national nonprofit network that was initially established with the assistance of federal grants.

Planning should begin now to develop a comparable system for obtaining and processing fetal tissue based on the framework of the originally proposed NIH fetal tissue bank. This would have been composed of a national network of collection, processing, and distribution centers at up to 20 institutions, each with its own network of subsidiary fetal tissue collection affiliates. Trained personnel would have processed fetal tissue, as is the case in the British fetal tissue bank. The original framework would have to be modified to include investigators and centers who respond to a recent request for applications (RFA) inviting proposals for studies of fetal tissue, including that derived from elective abortions (13). Centers chosen in the original grant competition should be continued only if their work is favorably evaluated and they fit into a rational framework of regional centers. Such a framework could be established by the NIH, which sets scientific and ethical standards for the rest of the country. A fetal tissue bank established under NIH auspices would carry out a large-scale, systematic comparison of the safety and suitability for transplantation of tissue from all major sources and would develop model standards for uniformity and quality control of fetal tissue that would be adopted around the country.

An initial issue to be considered is the administration of a fetal tissue bank. The Central Laboratory for Human Embryology at the University of Washington and the National Disease Research Interchange in Philadelphia, centers with a history of providing fetal tissue for medical research, have established criteria that could provide the starting point for developing administrative procedures for collecting, processing, and distributing fetal tissue. Other grantees and transplant investigators also could bring their experience and expertise to the development of methods for organizing the retrieval, distribution, and tracking of fetal tissue. Ultimately, a centrally coordinated national network of regional centers, along with transplant surgeons and other consultants, could develop administrative standards for equitable distribution of fetal tissue.

Secondly, the safety and suitability of the tissue to be

used in transplantation must be carefully monitored. Tissue derived from spontaneous abortions is often infected and bears a small risk of transmitting human immunodeficiency virus (HIV) and other potentially fatal viral infections (14). Tissue from elective abortions may be infected by vaginal flora or intrauterine transmission of maternal infections such as syphilis, herpes simplex, toxoplasma, chlamydia, cytomegalovirus, rubella, hepatitis B virus, and HIV (4, 14-18). Published reports from several American transplantation teams indicate that they are attempting to counter many of these risks. For example, tissue from elective abortions is rejected at the University of Colorado if either the donor or her sexual partner has a history of venereal disease, hepatitis, infection with HIV, or intravenous drug abuse (19). At Yale University, fetal brain tissue is screened for bacterial, fungal, viral, and mycoplasm a contamination, and donor serum is tested for HIV and hepatitis B (20).

More extensive information has been published about procedures used to evaluate and reduce the risk of infection from fetal tissue at the central fetal tissue bank in London (4). Fetal tissue from donors for whom an abortion was performed because of infection such as rubella virus, cytomegalovirus, HIV, hepatitis B, and toxoplasmosis is not used. Fetal tissue from donors who are known carriers of certain diseases, such as hepatitis B or HIV, also is avoided. The blood of donors and fetal tissue are tested for evidence of infection from syphilis, hepatitis B, and HIV. If the tissue is cryopreserved, the HIV-negative donor is retested after at least 90 days, and the tissue is used only if the donor is seronegative. Fetal tissue is washed in sterile solutions, and in antibiotics and antifungal solutions prior to transplantation. As a way to reduce the risk of transplanting genetically abnormal tissue, tissue from spontaneous abortions is not used if karyotypic or DNA analyses cannot be performed or if the fetus shows obvious anatomical abnormality. Because of the variability of tests currently performed at different laboratories, it would be advisable to develop some universal codification of testing procedures for fetal tissue, as is done in blood banking, to reduce infective and genetic risk.

A third matter to be addressed is the cost of such a bank. This depends not only on the amount of tissue needed, but also on the nature and scope of the screening performed. One rough estimate from A. Fantel, the director of the only laboratory in the United States with long-term experience at grading, separating, and analyzing fetal tissue, is that it would cost from $1000 to $2000 per specimen per year (21). This would cover donor interview and medical record examination; specimen procurement; grading, staging, and teratological examination; donor serologic screening for HIV and hepatitis viruses; specimen screening for mycoplasma and chlamydia; bacteriology; cytogenic examinations; histopathology; and the data compilation and review required to produce and deliver a complete report on each specimen. These costs include significant input from epidemiologic personnel as well as experimentation with tissue growth and cryopreservation. On the basis of this estimate, it can be concluded that to perform appropriate tests on the several thousand tissue specimens currently used for transplantation research per year would cost approximately $6 million.

Costs for administration and distribution would also need to be added to this estimate. The initial costs of establishing a national network of regional centers for fetal tissue banking should be borne, in part, by the federal government, in the way that the network of solid organ banks was established. The idea of profit from the sale of the human fetus or fetal tissue is incompatible with their special ethical significance (22). Therefore, regional centers should develop standards for nonprofit processing and distribution of fetal tissue.

Conclusion

There are good ethical and scientific reasons to retain a centrally coordinated fetal tissue bank system and to open this system to include tissue derived from elective abortions. Since fetal tissue transplantation promises to increase rapidly, this is the time to develop a set of ethical and scientific guidelines for such research in the United States. The fetal tissue bank, a once notorious political football, can become a primary source of standards for collecting and processing fetal tissue for transplantation. Moreover, it can provide a locus for developing equitable criteria for the distribution of this material. If this is to be accomplished, the scope and goals of the fetal tissue bank set out in the Executive Order of 1992 must be expanded, large-scale research must be carried out on tissue derived from elective abortions, and the bank must be adequately funded. ❑

References and Notes

1. Public Law 103-43, the National Institutes of Health Revitalization Act of 1993, 103 P. L. 43: 1993 S.1:107 Stat, 122 (103rd Congress, first session)
2. The five recipients were F. C. Zhou, Indiana University School of Medicine; W. C. Low, University of Minnesota; L. Ducat, National Disease Research Interchange, Philadelphia, PA; W. Branch, University of Utah; and A. G. Fantel, University of Washington.
3. M. L. Zoler, *Ob-Gyn News* 1, 24 (15 June 1993).
4. L. Wong, in *Fetal Tissue Transplants in Medicine*, R. G. Edwards, Ed. (Cambridge Univ. Press, Cambridge, 1992), p. 129.
5. D. P. Hamilton, *Science* 257, 869 (1992).
6. D. E. Vawter *et al., The Use of Human Fetal Tissue: Scientific, Ethical and Policy Concerns* (University of Minnesota, Minneapolis, 1990), p. 66.
7. M. Galdwell, *Washington Post* A3 (21 May 1992).
8. Committee to Review the Guidance on the Research Use of Fetuses and Fetal Material (The Polkinghome Report), *Review of the Guidance on the Research Use of Fetuses and Fetal Material* (Her Majesty's Stationery Office, London, 1989).
9. J. C. Polkinghome, in (4), p. 323.
10. *Report of the Human Fetal Tissue Transplantation Research Panel* (National Institutes of Health, Bethesda, MD, 1988).
11. L. Walters, Testimony on Human Fetal Tissue Research before the Subcommittee on Health and the Environment, Committee on Energy and Commerce, U.S. House of Representatives, 2 April 1990.
12. G. J. Annas and S. Elias, *N. Engl. J. Med.* 320, 1079 (1989).
13. RFA: HD-94-06, *NIH Guide* 22, 6 August 1993.
14. D. J. Goldberg *et al., Br. Med. J.* 304, 1082 (1992).
15. European Collaborative Study, *Lancet* 337, 253 (1991).
16. S. H. Lewis *et al., ibid.* 335, 565 (1990).
17. T. Oskarsson *et al., Acta Obstet. Gynecol. Scand.* 69, 635 (1990).
18. P. Levallois, J. Rioux, L. Cote, *Can. Med. Assoc. J.* 137, 33 (1987).
19. C. R. Freed *et al., N. Engl. J. Med.* 327, 1549 (1992).
20. D. D. Spencer *et al., ibid.,* p. 1541.
21. A. Fantel, personal communication.
22. C. S. Campbell, *Hastings Cent. Rep.* 22, 34 (1992).

23. Supported by the Ford Foundation, the Greenwall Foundation, the Walter and Elise Haas Fund, the Josiah Macy Jr. Foundation, and the Rockefeller Foundation. We thank L. B. Walters, Kennedy Institute of Ethics at Georgetown University, and J. F. Childress, Department of Religious Studies at the University of Virginia, for commenting on an earlier draft of this paper.

****

Questions:

1. What is the purpose of the fetal tissue bank?

2. Why should tissue from elective abortions be made available to the US fetal tissue bank?

3. Why must the tissue be rigorously tested before acceptance into the bank?

Answers are at the end of the book.

Irving Weissman's discovery has hit the scientific community like a hurricane leaving behind in its path a whirl of excitement and controversy. It all began when he staked his claim in the form of a patent on finding a human hematopoietic stem cell, a blood-forming cell capable of generating an endless supply of red cells, white cells, and platelets. Imagine the potential of such a monumental achievement as replacing a patient's entire blood supply in the treatment of cancers, AIDS and inherited diseases. The question is should one man and his colleagues be the proud owners of a living cell? Some of the scientific community believes otherwise since much of the work had already been done in other laboratories. Whatever the case, Weissman's patent has cast much interest and money into the field of stem cells and it stands to make big money for the winners.

The Mother of All Blood Cells

Stem cells, capable of generating an endless supply of red cells, white cells, and platelets, have also generated a heated scientific controversy — and millions of dollars for the man who claims to have found them.

by Peter Radetsky

Deep in the very marrow of our bones reside the living forebears of our blood, the hematopoietic ("blood forming") stem cells. From these rare and elusive pluripotent cells arise our oxygen-carrying red blood cells, the tiny platelets that facilitate coagulation, and the disease-fighting white cells of our immune system. The hematopoietic stem cells are nothing less than the springs that feed the river of life that flows through our veins. And Irving Weissman has found them.

To be precise, Weissman, a Stanford immunologist, and his collaborators at SyStemix, the Palo Alto, California, biotech company he cofounded in 1988, claim to have found a strong "candidate" for these remarkable cells. But they're not fooling anyone. So confident are they, that SyStemix has patented not only the process used to find the cells but the cells themselves, in effect claiming ownership of these biological entities. So confident is the giant Swiss drug-and-chemical company Sandoz Ltd., that it has bought 60 percent of the SyStemix stock for a reported $392 million, making the 55-year-old Weissman and his stockholders instant millionaires.

The stakes are indeed high. In addition to stem cells' importance in basic research, they may make possible a host of breakthrough medical advances. By giving patients the ability to make an entirely new blood supply essentially from scratch — and thus the ability to regenerate key components of the immune system — stem cell therapies could result in new treatments for various cancers, allow for bone marrow transplants without the need for rejection-fighting drugs, make possible powerful strategies against AIDS and other blood infections, and provide genetically engineered antidotes to a wide range of inherited diseases. No wonder Weissman has caused such a stir.

Yet none of the pioneering researchers in these fields currently employ the Weissman recipe for isolating stem cells — except, of course, those at SyStemix. Inarguably, the hunt for the stem cell preceded Weissman, and some say it would be proceeding just fine without him. While researchers agree that Weissman has brought recognition to a previously little-known field, they disagree as to the value of his scientific contribution. Some consider his work pivotal, others merely useful, others virtually irrelevant — regarding him, in Weissman's own words, as "a snake oil salesman." Despite Weissman's controversial patent, there is continuing debate as to exactly what stem cells really are and how they might best be tracked down. Says stem cell pioneer and Weissman competitor Malcolm Moore of the Memorial Sloan-Kettering Cancer Center in New York City, "No one has yet definitively isolated a stem cell. There's been a lot of talk, but it hasn't yet been pinned down so we can say, '*This* is the stem cell.'"

The modern search for the hematopoietic stem cell began with the detonation of the atomic bombs over Japan in 1945. Researchers could easily see that the intense blasts of radiation destroyed blood cells and that people often died within weeks of exposure. But scientists mimicking that exposure in mice soon realized that these deadly effects could be prevented by transplanting bone marrow from genetically identical donors into the irradiated mice. The injected marrow revived the irradiated blood; thus, the researchers reasoned, the marrow must contain cells capable of regenerating other blood cells, something that mature blood cells are not able to do.

In the 1960s those speculations became fact. James Till and Ernest McCulloch of the Ontario Cancer Institute in Toronto found that after bone marrow cells were injected into irradiated mice, the animals developed nodules on the spleen. Each nodule was chock-full of white and red blood cells. By tracking genetic markers in the cells' chromosomes, Till and McCulloch saw that the cells within each nodule had all derived from a single progenitor: one per nodule. Then, by simply counting nodules, they were able to estimate the number of progenitor cells in each batch of transplanted marrow. The cells turned out to be rare, about 1 in 1,000. Furthermore, Till and McCulloch found that in addition to generating a wide range of new blood cells, these progenitor cells were also able to reproduce themselves.

Based on this evidence, Till McCulloch came up with a scenario that has been considered gospel ever since. All blood cells, they said, arise from a few hematopoietic stem cells, which are hidden away in the bone marrow. (Blood-forming stem cells are not the only ones we harbor: there are supposedly stem cells for skin, liver, and the intestines, as well as stem cells behind the generation of eggs and sperm). These cells, as remarkable as they are rare, can both renew themselves and produce trillions upon trillions of blood cells, an inexhaustible supply for the life of their host body. When these new cells mature and die off — human red blood cells, for example, last only 120 days — the stem cells produce more to take their place. On average these cells produce an ounce of new blood — some 260 billion new cells — each and every day.

Irv Weissman was a Stanford medical student and research associate when Till and McCulloch did their groundbreaking work. "I knew those experiments," he says. "I knew them cold. They were thrilling." The experiments have gave him some insight into the problems of organ transplantation, something he'd been interested in since his high school years in Great Falls, Montana. "I thought that the most important thing would be to understand the development of the immune system," he says. "If you understood that, then when you did transplants you'd know what was going on." Till and McCulloch's stem cell revelations gave him a path to follow. I began getting further and further into understanding white cell development, moving backward from mature cells to earlier and earlier cells."

Of course, Weissman wasn't the only one moving backward. Researchers all over the world were beginning to look at the early, immature blood cells. Among them were the hematologists and biologists who made up the "Dutch Mafia," as biophysicist Jan Visser laughingly describes himself and his compatriots. And it was the Dutch Mafia that first hit stem cell pay dirt. In 1984 Visser announced that he and his colleagues in the Netherlands had isolated stem cells in mice.

It was a startling achievement, one that eluded even Till and McCulloch. The Ontario researchers had only been able to document the cells' existence, not to pin them down. But in the intervening years molecular techniques had grown more sophisticated. Visser had at his disposal molecular probes designed to find their prey by homing in on any number of unique characteristics. The task of finding the theorized stem cells among all the varied cells in a sample of marrow was therefore akin to trying to pick someone out of a crowd by looking for a particular combination of hair color, weight, and nose shape. This didn't mean the job was an easy one, however. Stem cells are comparatively primitive cells that lack the diversity of features associated with their mature progeny; they seem to be distinguished primarily by their dearth of unique characteristics.

To tease out the stem cells, then, Visser employed a three-pronged strategy. First he separated the cells by density. It had been discovered some years earlier that cells with stem cell activity — that is, cells that gave every indication of indeed being the long-sought hematopoietic cells — tend to be lower in density than other bone marrow cells, so Visser placed a batch of bone marrow cells (between 60 million and 100 million of them) in a centrifuge and culled only the ones that rose to the surface. That one step got rid of some 90 percent of all the cells.

Next he turned to a substance commonly used in laboratories to purify proteins: wheat germ agglutinin. Agglutinin fuses with certain sugars associated with proteins, and Visser had found that it also sticks to the sugars in the membranes of cells with stem cell activity. So he took the remaining 10 percent of the cells and mixed them with wheat germ agglutinin tagged with a fluorescent dye. With the help of a fluorescence-activated cell sorter, he was able to separate out only those cells that emitted a fluorescent glow, a sign that the tagged agglutinin was holding tight. In this way Visser further reduced his sample by 90 percent. He was now left with just 1 percent of his original bone marrow mix.

Finally he employed monoclonal antibodies. Antibodies are large Y-shaped molecules that are among the immune system's prime infection fighters. They make a beeline for foreign proteins, grab them, and mark them for destruction by other immune forces. By the early 1980s these tiny guided missiles were among the favorite tools of molecular biologists, since they could be engineered to go after almost any target a scientist might choose. Visser sent them after one of the few stem cell characteristics then known: a protein called H-2K, which he had discovered in greater numbers on the surfaces of stem cells than on any other cell. Cells the antibodies ignored couldn't have the protein and thus couldn't be stem cells. These he cast aside.

The result was a further narrowing of the search. The antibodies reduced the number of cells by two-thirds: what remained was just three-thousandths of the original blend, only about 200,000 cells. When Visser injected these relatively few cells into irradiated mice, he found that it took no more than 200 of them to regenerate each animal's entire blood system. In other words, there was at least 1 stem cell in every 1,000 bone marrow cells, the very proportion Till and McCulloch had come up with. Of course, Visser's sample wasn't pure — there were other cells in the mix — but out of an initial crowd of tens of millions, he was pretty close. Three years later, using newer sorting techniques, he whittled it down even further, using only 30 cells to save an irradiated mouse. He now estimates that stem cells represent 1 in 10,000 marrow cells.

Visser published his original results in the *Journal of Experiment Medicine* in 1984. Fours years later Weissman announced in the journal *Science* that he and his colleagues had found the mouse stem cell. Whereas Visser's work had elicited polite praise, Weissman's made headlines. "The *Journal of Experimental Medicine* is considered scientifically one of the best journals," Visser notes with some irony. "*Science* is more popular."

Weissman and his team had taken a much more narrow approach to ferreting out their prey: they relied solely on a variety of monoclonal antibodies, each designed to pick out a different stem cell surface protein — or proteins on other cells that they had found to be absent on stem cells. For instance, one group of monoclonal antibodies targeted proteins found only on the surface of mature bone marrow cells, thus allowing the researchers to get rid of almost everything that was not a stem cell. Weissman dubbed the remaining cells Lin^- ('lineage minus") because the antibodies had subtracted all other cell lineages. "We used those antibodies to get rid of 90 percent of the bone marrow," he says. "The 10 percent that was left had stem cell activity."

To pinpoint the source of that activity more precisely, they used two other monoclonal antibodies targeted to two surface proteins they knew appeared on stem cells. One — Thy1 — is found in low concentrations (designated "lo"); the other — Sca1 — is far more common (designed "+"). By throwing out the cells that did not display Thy1 or Sca1, Weissman brought the number of remaining cells down to just .05 percent of the whole. He dubbed the cells left behind $Thy1^{lo}Lin^-Sca1^+$; like Visser, he found that only 30 of them were needed to reconstitute the full range of blood cells in an irradiated mouse. What he had was a "virtually pure" batch of stem cells.

Reaction was strong and swift. *Science* accompanied Weissman's paper with a news story entitled "Blood-Forming Stem Cells Purified," in which Weissman, without so much as a word acknowledging earlier work toward the same goal, was quoted as saying, "This is the end of the particular road that was the search for the stem cell." He contended that his methodology — based on identifying the actual look of the cells — was so efficient that it might be used to go after *human* stem cells. Neither Visser — who had focused more on the cell's density and their propensity to bind to agglutinin — nor anyone else had made a claim like that.

SCIENTISTS CLOSE IN ON A VITAL BLOOD CELL, proclaimed the *Wall Street Journal.* THRILLED TO THE MARROW: BIOLOGISTS FINALLY CORNER THE RARE FOREBEAR OF ALL BLOOD CELLS, announced *Scientific American.* Weissman spread the word over television and radio, and the news was covered worldwide by the Voice of America. But despite the public applause, not all the response from the scientific community was complimentary. An editorial in *Immunology Today*, for example, while acknowledging that the researchers had indeed isolated a cell population with stem cell activity, asked, "But does this represent any advance on previously published data?" The editorial pointed out that Visser's work had "generated populations with similar characteristics and only moderately less purity," and concluded, guardedly, "We have not yet reached the end of the road; perhaps just one of the side streets."

The suspicion spread that Weissman had done little that was new but had nevertheless claimed credit for a monumental discovery. In the process he had slighted the achievements of earlier pioneers. Visser's in particular. Whether Weissman's greater recognition was the result of his being published in the right place at the right time, or his being American rather than Dutch, or his simply knowing his way around a press conference better than Visser, many scientists felt that Weissman was receiving attention out of proportion to his accomplishment.

"Did the Weissman group make a real contribution in the mouse? No, not at all," declares Malcolm Moore. "All of the work had already been done elsewhere. Visser had done it a long time before. So there were some people very, very angry that he had taken credit for discovering the stem cell when he hadn't done any of the primary work."

"It's unfortunate that it worked out that way," says stem cell researcher Ihor Lemischka of Princeton. "I think Weissman would be the first to admit that his getting all the credit was unfair. It caused a lot of bad feeling. Now there are two camps in the field. There's the Dutch axis and the Weissman crowd, and everybody else is in between."

Weissman has since explicitly acknowledged Visser's prior contribution. He didn't do so at the time, he says, because he simply wasn't aware of Visser's work. "I didn't even know about it until we finished our work," Weissman claims. "Early in my career a prominent scientist told me that when he started something in earnest, he quit reading the literature. He didn't want to be distracted by what others were doing, or spoil the fun of discovery. I use that as an excuse. It's a bad thing to admit this, I know."

Whatever its cause, Weissman's omission had unfortunate consequences. Says Moore, "Weissman was an immunologist who suddenly hit the interface with experimental hematology without appreciating that people had been working for many, many years and knew about all these things. But somehow he obtained the credit for discovering stem cells. He trod on many toes."

Ten of those toes belonged to Visser, today the head of the recently formed stem cell laboratory at the New York Blood Center in Manhattan. Yet Visser is far less strident in his assessment of Weissman's tactics than are some of his colleagues. In fact, he seems almost to admire Weissman's flair. "Did I feel that Weissman stole my thunder?" he muses. "At first I did. I felt sort of sorry about it. But that soon went away. He raised attention to this field that I could never have raised. In his jet stream, in all the noise he makes, I was drawn with him. We both traveled to meetings around the world to explain our differences. No, after all it was no problem. It was good for me. It gave me a lot of attention."

Whether he was the first to

find stem cells, it is undeniable that Weissman devised a very useful approach. It was precisely because he was an outsider that he was able to pick and choose among existing techniques and combine them in a fresh way. $Thy1^{lo}Lin^{-}Sca1^{+}$ was an original recipe for a stem cell.

"The real value of the Weissman purification is that it describes cell-surface differences," says Lemischka. "Thy1, Sca1, and lineage markers — these are tangible molecules. So in terms of saying, 'A stem cell looks like this, it has on its surface this and that, but not these,' Weissman defined the stem cell for the first time. Weissman's accomplishment is more than just a purification like Visser's; it is really informative."

McCulloch is even stronger in his praise. "Weissman's remains the most extensive purification that's been achieved," he says. "I consider his work to be seminal. The application to man was just a step beyond what he did in the mouse."

All the same, it was quite a momentous step. By the winter of 1991 Weissman and his colleagues at SyStemix had adapted the mouse stem cell recipe and announced that they had found the human stem cell. As with the mice, they selected against the lineage markers, and for the Thy1 marker. Sca1 wasn't a practical marker to use in human cells — in humans the gene that encodes the protein has many similar-looking relatives — so the researchers substituted a protein called CD34, which had been discovered some years earlier. They dubbed the human stem cell $Thy1^{+}Lin^{-}CD34^{+}$. Once again Weissman's individual ingredients weren't new, but his combination was. Yet once again the scientific community wasn't overawed — other researchers, using many of the same markers, were already finding similar cells.

On the other hand, none of those other researchers announced their achievement the way Weissman did — with a patent. In a dramatic departure from normal scientific procedure, Weissman and his colleagues patented their method for finding their purported human stem cell *and* the stem cell itself some five months before their work appeared in the April 1992 issue of the *Proceedings of the National Academy of Sciences.* Since a patent confers exclusive rights to the thing patented, this meant that Weissman and his colleagues were the proud owners of a living human cell.

It was an audacious stroke, if not an entirely unprecedented one. "You know, people patent body components all the time," says Weissman. He's right. Genes, growth factors, blood proteins — all have received patents. Even CD34 is patented — SyStemix had to pay to use the protein. But an entire living cell?

"Next thing you know, someone will patent a zygote, and you won't be able to have a baby without a license," quipped Sloan-Kettering hematologist David Golde at the time of the announcement."

If somebody wants to commercialize having babies, who knows?" replies Weissman. As with any patent, SyStemix's patent is specifically designed to forestall the commercial exploitation of stem cells by anyone other than itself. And the company intends to guard its privilege aggressively. "It doesn't matter what process people use, they will always be infringing," declared SyStemix's then president, Linda Sonntag.

Such talk infuriates Moore and others in the field. "It seems to me an absurd patent," Moore says. "I predict it won't withstand any challenge. If they think that anybody who tries to separate and grow stem cells is going to have to pay a licensing fee to SyStemix, they've got another think coming. Because people can separate stem cells using different criteria from the ones outlined by the SyStemix patent." For example, in January, Harvard Medical School researchers announced that they had developed an entirely novel and relatively simple method of isolating stem cells based not on antibodies but on the principle that stem cells, which spend most of their time in a quiescent, nondividing "deep sleep," are relatively unresponsive to proteins known as growth factors. The stem cells' progeny, on the other hand, do respond to growth factors. So the Harvard team, led by hematologist David Scadden, turned on the stem cells' offspring by applying those factors and then promptly killed them by smothering the same cells with an anticancer drug called 5-FU, which attacks metabolically active cells. Left behind were stem cells.

Jan Visser is much more philosophical than Moore. Visser sees the move as a practical necessity. "Weissman needed the patent for his company."

It's an analysis with which Weissman agrees. "Money people won't invest in something that's not patented," he says. "And if you don't get money, you can't do big-time research."

Weissman can do that research. Almost immediately after the patent announcement, Sandoz bought controlling interest in SyStemix, ballooning Weissman's net worth, at the time, to an estimated $24 million. "It hasn't changed my life," he says, then chuckles. "Well, I drink better wines than I did."

"I would have liked to be as rich as he is," says Visser wistfully. "It's my own mistake. He did the right thing — from a family point of view." He laughs. "My children are complaining."

In the clinical realm, however, the SyStemix patent has yet to become a real factor. In its own labs, SyStemix has shown that its human stem cells can indeed regenerate blood in mice that have been bred with a human immune system — the so-called SCID-humice. And the company has received permission to begin testing its ability to do so in humans by injecting the cells into 20 multiple myeloma patients whose immune systems have been ravaged by chemotherapy. These tests should be starting right around now; it's taken this long to develop a technology that can use the Thy1+Lin-CD34+ recipe to isolate stem cells in large enough quantities to use in humans.

But none of the other clinical trials going on right now use the Weissman recipe. At the New York Blood Center, immunogeneticist Pablo Rubinstein is spearheading a worldwide effort to transplant stem cells by utilizing human neonatal cord blood. His approach is simplicity itself. In the fetus, stem cells continue flowing through the blood before settling into the bone marrow a few days after birth. Cord blood is therefore relatively rich in stem cells; Rubinstein and his colleagues can simply collect and transplant that whole blood and know that they are transplanting stem cells. Since the fall of 1993 the team has helped perform transplants on 13 patients — all but one of them children — suffering advanced stages of leukemia or inherited disease. The idea is to provide these people with a new source of blood to replace their own diseased cells. The results have given researchers grounds for hope. Although 3 of the patients (including the adult) died from unrelated complications, the other 10 are doing well. Their blood has been fully regenerated, their disease put into remission.

Moreover, none of these patients experienced the bane of transplantation efforts, graft-versus-host disease, in which the introduced cells recognize the host's body as foreign and attack it. "With pure stem cells, there should be no graft-versus-host disease," says Rubinstein. "The immune system generated from such a cell would mature in the recipient and thus would be trained to recognize the recipient as self."

Another trial using cord blood began at Children's Hospital in Los Angeles in the spring of 1993, when pediatric gene therapist Donald Kohn performed stem cell gene therapy on three newborns afflicted with an inherited immune system disorder called ADA deficiency, known more popularly as the bubble boy disease. Children with the disease lack the gene to produce the enzyme adenosine deaminase, or ADA. Without ADA the immune system cannot function; such children face certain early death.

ADA deficiency was the target of the very first gene therapy trials, begun in 1990 at the National Institutes of Health. Those trials involved putting corrected genes into mature white cells rather than stem cells. The corrected cells thus eventually die, which means that the therapy — while revolutionary and successful — will be able only to curtail, not cure, the disease. Stem cells, on the other hand, might provide a true cure. In his attempt to provide that cure, Kohn actually needed to isolate the stem cells in the babies' cord blood: he did so by targeting the CD34 protein, inserting the ADA gene in the cells, and returning them to the infants. He's hopeful that the genes have taken up residence in the stem cells and are being packaged into their progeny. "For the first few months we didn't see any cells with the gene," he says. "Then we started seeing a few, maybe 1 in 10,000. Now we're up to about 1 in 1,000." Although in the meantime the infants receiving injected doses of ADA enzyme, Kohn's hope is that stem-cell-generated cells containing the inserted gene will eventually produce enough enzyme to obviate the need for supplemental therapy and effectively cure the children.

These trailblazing efforts may presage a spectacular future for stem cell therapies. Malcolm Moore and his team are inserting drug-resistant genes into stem cells identified by CD34 and other markers so that they can give the cells to leukemia patients or cancer patients undergoing chemotherapy, which tends to destroy healthy white cells along with the diseased cells. A number of groups — including one at SyStemix — are looking at inserting anti-HIV genes into stem cells as a possible treatment for AIDS. Finally, though it's a long shot, the ability to isolate stem cells may even lessen the escalating need for blood transfusions.

In the end, controversy and hard feelings aside, it seems that Weissman has sparked a revolution. Today there is interest in stem cells, and money available to fuel that interest, as never before. "Because of the public relations work that Weissman did, it became easier to get grants," says Visser. "Because Sandoz bought SyStemix, all the big companies started to look at stem cells."

"What did Irv Weissman bring?" ask stem cell researcher Norman Iscove of the Ontario Cancer Institute. "He made the field live. He brought it into headlines, into newscasts."

No one appreciates the field more than Weissman. "I think you'll see the first practical large-scale use of stem cells in three years for cancers, five to seven years for other conditions," he says. "The great thing is that we have developed this population of cells that are just the right thing. They're not like drugs, which have side effects. You know that what you're providing is the right thing because stem cells are the product of over a billion years of evolution."

And Irv Weissman owns them. ❑

Questions:

1. What are stem cells?

2. How did Irv Weissman discover stem cells?

3. How can stem cells be used for therapy?

Answers are at the end of the book.

RU486 (Mifepristone) is widely known as the French "abortion pill" for its ability to terminate unwanted pregnancies by preventing the effects of the hormone progesterone. The drug mifepristone, however, is more than an abortion pill to some but a promising treatment for breast cancer and brain tumors. The drug may also be useful in the treatment of endometriosis, uterine fibroids and Cushing's syndrome. If it has so much potential in human health, why aren't researchers able to get it? Part of the reason is that no US manufacturer has been named due to the public's opposition to the drug. Despite the political pressure to keep the drug out of US, RU486 will be submitted to the FDA for approval in 1996.

Researchers face delay in supplies of RU486

by Helen Gavaghan

Washington. The politics of abortion in the United States is holding up clinical and preclinical research on RU486, the so-called abortion pill, which also promises many other biomedical applications in treatments ranging from breast cancer to brain tumours.

Earlier this year, faced with the heat surrounding the abortion debate, Roussel-Uclaf, the manufacturers of the drug, handed over the US patent rights to the Population Council in New York. The council, a nonprofit group, has agreed to find a manufacturer and distributor in the United States and make the required New Drug Application to the Food and Drug Administration.

But the council — which has yet to name a manufacturer — will not have a stock of the drug for at least a year. Meanwhile, Roussel-Uclaf has stopped supplying the drug to new investigators. Supplies to existing trials, such as those for a type of brain tumour and the Population Council's abortion trials, will continue.

Wayne Bardin, director of the council's Center for Biomedical Research, has heard from about 20 investigators who have been affected by supply problems; some began writing to him in September to ask for supplies of the drug at the suggestion of Roussel-Uclaf. Yet despite such demand, Andre Ulmann, director of research and development in endocrinology at Roussel-Uclaf, says the company will not be supplying RU486 to new investigators because of its contract with the Population Council.

One possible solution suggested by Bardin is for the Population Council to distribute the drug for Roussel-Uclaf until stocks are manufactured in the United States.

The current situation is the latest manifestation of the problems faced by US scientists interested in research with RU486, RU486 (mifepristone), is an antiprogestin, and it is important because although it binds progesterone receptors it has little progesterone-like activity. It prevents the progesterone effects needed to establish pregnancy, thereby causing abortion and it is this property that has created the political pressure to keep it out of the US.

But, as an antiprogestin, RU486 has much wider possible applications. In 1993, the US Institute of Medicine (IOM) released a report on the clinical applications of antiprogestins, describing them as having "a significant potential in human health". The institute cited the need to investigate the potential use of antiprogestins as a therapy for meningioma (a tumour of the membrane surrounding the brain), endometriosis, breast cancer and uterine leiomyomas (fibroids).

Mary Lake Polan, chair of the department of obstetrics and gynecology at Stanford University in California, and a member of the committee that prepared the IOM report, says that in the current climate it is difficult to blame Roussel-Uclaf for wanting to keep RU486 at arms length.

Polan describes the IOM report as "an attempt to put RU486 back into the realm of science". But the move does not seem to have succeeded. Several scientists failed to return repeated telephone calls from *Nature*. "You may have noticed [anti-abortionists are] shooting people," said one who did. "It's no longer a science issue, and it's hard to predict what people driven by a conviction of their own moral superiority will do."

One scientist who has repeatedly spoken out on the potential value of RU486 is Sam Yen, professor of reproductive medicine at the University of California, San Diego. "The emerging evidence is that this powerful new drug might have utility in the brain and the reproductive system," says Yen. "But these potential benefits are being deterred by politics." His own work was halted 18 months ago when he was caught up in a battle between Roussel-Uclaf and its German parent company Hoechst over whether to continue producing RU486. But he too is reluctant to blame the company when "their property might be burnt".

Yen has recently received a supply of RU486 through an informal international group of scientists. This source, however, is unable to meet all demands for the drug.

Canadian and European researchers are already evaluating RU486 for treating breast cancers. But the US National Cancer Institute has decided not to pursue research with this molecule. "We watch the literature, but for now we have many other promising therapies," says Michaele Christian, head of the National Cancer Institute's developmental chemotherapy section.

Another part of the US National Institutes of Health, the National Institute of Child Health and Human Development, has sponsored work at the Research Triangle Institute in North Carolina to develop an antiprogestin. Ed Cook, the head of the project, says it has a candidate compound but that it is a long way from being available for research with patients.

Roussel-Uclaf continues to face controversy, with both external and (it is said) internal pressures to abandon RU486 completely. Ulmann admits there are rumours that the company is pulling out of work on antiprogestins, but says that no such decision has been made. ❑

****

Questions:

1. Why is RU486 called the "abortion pill?"

2. What are some of the potential uses of RU486 in human health?

3. What is causing the delay of RU486 supplies in the US?

Answers are at the end of the book.

Answer Section

PART ONE: LIFE IN THE CELL

-01- Ringing Necks with Dynamin

1. Synaptic vesicles, found in the synaptic knob of neurons, are pouches which store specific neurotransmitters for release each time a neuron fires.

2. Endocytosis is the process of internalization of extracellular material within a cell.

3. Dynamin is a large protein whose helical structure wraps around the neck of membrane vesicles and helps to pinch them off from the plasma membrane.

-02- Teaching Tolerance to T Cells

1. An autoimmune disease results when the body's own immune system fails to recognize self proteins from nonself proteins and attacks the body's own tissues.

2. By feeding synthetic protein to the body, it is possible to reduce the actual number of attacking T cells and suppress the activation of harmful T cells involved in the immune response.

3. Oral tolerance could never eliminate autoimmune disease but only restrain the immune response.

-03- Presenting an Odd Autoantigen

1. In MS, T cells, B cells and macrophages attack the myelin sheath that surrounds nerve fibers.

2. This protein is expressed in glial cells from lesions in the brain of MS patients.

3. It is possible to induce immunological tolerance to this protein.

-04- Garbage Trucks of the Cell

1. Cellular proteins responsible for clearing the cell of unwanted waste proteins which accumulate with time.

2. It grinds the waste into small peptide fragments.

3. Specialized proteins guard the entrance of the proteasome which selectively unfolds proteins marked with ubiquitin.

-05- Cell Biologists Explore 'Tiny Caves'

1. Caveolae are small indentations on the cell membrane surface which draw substances into the pit to be released into the cell's interior.

2. Caveolin, a marker protein, covers the pits in caveolae and is used by scientists to identify the caveolae.

3. Transcytosis is a process in which molecules are picked up and transported across endothelial cells to the membrane on the other side where their contents can be released into the fluids bathing the tissues.

-06- Crucibles in the Cell

1. Once it engulfs the antigen, the B cell breaks down the invader's proteins into peptides.

2. The MHC molecules are distinctively yours and inform the immune system to attack.

3. The CPL, compartment for peptide loading, forms a complex with self and nonself proteins which gets escorted to the membrane as a signal to the immune system's attack cells.

-07- Revisiting the Fluid Mosaic Model of Membranes

1. The fluid mosaic model describes the plasma membrane as a phospholipid bilayer with proteins associated with either the inside, outside or spanning the membrane completely in an unrestricted manner.

2. Proteins confined transiently within the membrane are the cell adhesion molecules, neural adhesion molecules and epidermal growth factor receptors.

3. The membrane-spanning proteins are sterically confined to the cytoplasmic face of the membrane face based on their cytoskeletal mesh size.

-08- Bicarbonate Briefly CO_2-Free

1. Carbon dioxide and water form carbonic acid which dissociates into H^+ and bicarbonate.

2. Carbonic anhydrase catalyzes the reaction of CO_2 with water.

3. Potassium creates a gradient necessary for the cotransport of bicarbonate in this new system.

PART TWO: WHAT'S NEW IN TISSUE ENGINEERING

-09- Old Protein Provides New Clue to Nerve Regeneration Puzzle

1. Myelin-associated glycoprotein or MAG

2. MAG is found in the fatty myelin sheath that surrounds nerve fibers in both the central and peripheral nervous systems.

3. MAG is found in lesser quantities in the peripheral nervous system and probably not around at all when they regrow.

-10- Neurotrophic Factors Enter the Clinic

1. Neurotrophic factors are naturally occurring proteins that keep neurons alive and healthy during embryonic development and later in normal adult life.

2. Neurotrophic factors can improve surviving neurons by stimulating them to grow more projections.

3. For the treatment of Alzheimer's and Parkinson's diseases, the trophic factors must be delivered directly to the brain since they would never pass the blood-brain barrier via the bloodstream.

-11- Tissue Engineering

1. In tissue engineering, live cells are seeded into gossamer sheets of dissolvable material and allowed to incubate in a bioreactor until they form a tissue suitable for transplant.

2. Hormone-secreting cells such as pancreatic cells could be injected into the abdomens of people with diabetes, thereby releasing insulin as needed.

3. Growing tissue from a patient's own cells would help to avoid an immune reaction.

-12- Researchers Broaden the Attack on Parkinson's Disease

1. Graft survival is quite variable in transplanted tissue.

2. It would produce almost pure dopaminergic neuroblasts and reduce or eliminate the dependence on fetuses.

3. It is a protein that does not cross the blood-brain barrier.

-13- Fetal Attraction

1. Substantia nigra, the region that supplies the brain with the neurotransmitter dopamine, is destroyed.

2. L-dopa, or levodopa, is a drug that can pass through the blood-brain barrier to be converted to dopamine.

3. Transplants from the patient's own tissue.

PART THREE: THE BRAIN, THE PERIPHERAL NERVOUS SYSTEM AND PROPERTIES OF NEURONS THAT SERVE THEM BOTH

-14- Remapping the Motor Cortex

1. Primary motor cortex.

2. Most neurons connect to a number of different muscles.

3. A dispersed system with extensive neural networks would allow constant formation of motor neuron connections for movement control, the kind associated with learning new motor skills.

-15- Helping Neurons Find Their Way

1. Axons are the nerve cell projections that make connections with their target cells.

2. There are short-range repulsive and attractive cues provided by molecules on the surface of nerve cells and diffusible chemoattractants and chemorepellents which act over longer distances.

3. Netrin-1, a guidance protein, is made by the floor plate, a strip of tissue that runs along the lower surface of the developing spinal cord.

-16- Brain Changes Linked to Phantom-Limb Pain

1. Phantom-limb pain is described by amputees as the feeling that pain exists in the phantom limb at specific anatomical points.

2. Somatosensory cortex

3. Sensory sites for other body areas will expand into the region previously owned by the amputated limb.

-17- Interfering With the Runners

1. ALS is a motor-neuron disease characterized by the degeneration of descending pathways in the large motor neurons of the body.

2. Scientists are using transgenic animal models to simulate the ALS condition and help illuminate the cause of the disease at the molecular level.

3. The accumulation of neurofilaments in the proximal axon prevents the delivery of energy-rich mitochondria to the axon causing death to the neuron.

-18- Wallpaper for the Mind

1. Scientists discovered that depth perception arises from the way the brain compares signals from both the eyes.

2. The brain perceives the 3-D first and the object recognition comes later.

3. Object-recognition and depth-perception regions of the brain may be working in tandem, bouncing signals back and forth.

-19- Dendrites Shed Their Dull Image

1. Each synaptic potential adds up with all the others moving through the dendrites, which if great enough, will result in an action potential.

2. In the patch-clamp technique, electrodes are pressed against the neuron's outer membrane to form a tight electrical seal so that action potentials can be measured at the dendrite.

3. Backward-traveling of signals can strengthen the synapse (plasticity) which will allow it produce a better response to future signals.

-20- Keeping Synapses Up to Speed

1. An action potential caused by a depolarization of a nerve terminal can initiate exocytosis.

2. Synaptic vesicles are a special form of endosome-derived vesicles which carry neurotransmitters to the synaptic membrane to be released during neurotransmission.

3. Synapsin helps to ensure a steady supply of fusion-competent vesicles ready for release at synaptic sites.

-21- Learning by Diffusion: Nitric Oxide May Spread Memories

1. Long-term potentiation is the neural mechanism which may be responsible for memory formation. It occurs when impulses across the synapse of two neurons results in a positive feedback effect, making future transmissions easier.

2. Nitric oxide is a highly reactive soluble gas.

3. They applied a chemical that specifically blocked the production of nitric oxide in a postsynaptic neuron, thereby causing long-term potentiation to fail in that synapse and in neighboring synapses.

-22- More Jobs for that Molecule

1. Nitric oxide is found in several places in the body; it is located in the endothelial cells of blood vessels, produced by macrophages after endotoxin invasion, and released as a neurotransmitter in the brain and peripheral autonomic nervous system.

2. They study nitric oxide synthase, the enzyme that makes nitric oxide.

3. Nitric oxide has been identified in skeletal muscle where it is concentrated in the fast fibers.

-23- S/He Brains

1. Functional magnetic resonance imaging was used to study the brain and detect shifts in blood flow that indicates which parts of the brain are more active than others.

2. The inferior frontal gyros of the left hemisphere showed an increased blood flow in both sexes, however, the brains in women also showed some activation in the corresponding region of the right hemisphere.

3. Knowing that women process letters into sounds using both sides of the brain may help explain some important observations like why women tend to be more successful at overcoming language disabilities.

-24- Seeing the Mind

1. Positron-emission tomography and magnetic resonance imaging.

2. They image aspects of blood vessel function that reflect neuronal activity.

3. Scientists will be able to identify the individual differences in brain anatomy as it relates to mental function.

-25- Serotonin, Motor Activity and Depression-Related Disorders

1. Serotonin is a neurotransmitter.

2. To prime and facilitate gross motor output.

3. Drugs, like Prozac, prevent a neuron from taking serotonin back into the cell which results in a higher level of activity.

PART FOUR: THE MECHANICAL FORCES AND CHEMISTRY OF THE SENSES

-26- Mechanoreceptive Membrane Channels

1. Phasic and tonic receptors
2. Mechanical stimulation leads to a transient increase in a receptor membrane's conductance to the ions sodium, potassium and sometimes calcium.
3. Certain drugs can block the channels for study in various types of cells.

-27- Neurons Tap Out a Code That May Help Locate Sounds

1. The auditory cortex is the area of the brain responsible for sound perception.
2. Neurons in the auditory cortex change their temporal pattern of firing as a function of sound location.
3. The ears are mapping the frequency of the sound and not the location.

-28- The Smell Files

1. Within the olfactory bulbs are tiny glomeruli which act as neural junctions to sort the signals coming in from the receptors about odor components.
2. Signals don't get mixed up because only neurons bearing one kind of receptor converge on a single glomerulus.
3. An odor is distinguished by the pattern of glomeruli it activates.

-29- The Sense of Taste

1. Salt, sour, sweet and bitter
2. Taste buds
3. G-proteins

-30- A Taste of Things to Come

1. Salty, sour, sweet and bitter

2. Each taste is mediated in some way by signalling pathways that result in receptor cell depolarization, release of neurotransmitter and synaptic activity.

3. A Gα subunit that is highly specific to taste neurons and used in the bitter signaling cascade.

-31- Envisioning an Artificial Retina

1. The light receptors of the eye are the rods and cones of the retina.

2. The ganglion cells will receive information from the implant on the inside of the retina where they feed into the optic nerve.

3. Electrical activity from the implant will need to translate visual stimuli to the visual cortex in the brain.

PART FIVE: IMPORTANT MESSENGERS OF THE RENAL, ENDOCRINE, CARDIOVASCULAR, AND REPRODUCTIVE SYSTEMS

-32- Thyroid Diseases

1. The thyroid regulates the body's basal metabolism.
2. Thyroid-stimulating hormone, released from the pituitary, signals the thyroid to produce more T3 and T4 when needed.
3. Hypothyroidism

-33- Drug of Darkness

1. Melatonin is a hormone secreted at night by the pineal gland within the center of the brain.
2. Melatonin suppresses ovulation.
3. Melatonin can act as a scavenger of free radicals - molecules that possess an unpaired electron.

-34- Assisted Reproduction

1. FSH, follicle-stimulating hormone, is produced by the pituitary gland in an effort to get the ovary to release an egg.
2. 10-15%
3. All have similar pregnancy rates, averaging 18-19%, but vary between clinics.

-35- Forever Smart

1. Estrogen is a hormone produced by the ovaries.
2. Menopause begins when estrogen production by the ovaries declines to almost nothing.
3. Women take estrogen at menopause to combat the undesirable effects such as osteoporosis and hot flashes.

-36- Not by Testosterone Alone

1. SRY gene
2. MIS actively suppresses the development of the female reproductive tract.
3. The MIS gene is found on the X chromosome so both males and females have it.

-37- Hormone Mimics Fabled Fountain of Youth

1. Adrenal glands

2. DHEA slowly declines with age.

3. Volunteers were given a dose of hormone to equal the amount found in the average 30-year-old.

-38- Exercised-Induced Renal and Electrolyte Changes

1. Plasma volume and electrolyte stores are depleted due to the loss of water and sodium in sweat.

2. Sweat loss can exceed 2 1/hr.

3. The primary ingredients are water and sugar in the form of corn syrup. Some sodium and other electrolytes can be found, as well as water-soluble vitamins.

PART SIX: HEALTH AND THE ENVIRONMENT

-39- Additional Source of Dietary 'Estrogens'

1. Exposure to estrogenlike pollutants is linked to breast cancer and reproductive abnormalities.

2. BPA is leached out of certain plastics when heated at very high temperatures.

3. Compared with estradiol, the major estrogen in humans, BPA is a thousand times less potent.

-40- Another DDT Connection

1. Androgens play a key role in the development of the male reproductive tract and in the masculinization of the body during fetal life.

2. DDT was banned for its persistence and accumulation in the environment and for its adverse effects on wildlife.

3. 100 years

-41- Beyond Estrogens

1. They block the activity of male sex hormones.

2. DDE readily binds to the androgen receptor.

3. Not much is known about DDE's effect on the cellular and genetic level.

-42- How Much Exercise is Enough

1. They selected a program that would benefit the greatest number of people and would be readily accepted.

2. Aerobic exercise is defined as activity intense enough to produce perspiration.

3. Weight-bearing exercises are necessary to preserve bone and reduce the risk of osteoporosis.

-43- Fixing Hemophilia

1. The blood is unable to form fibrous clots.

2. The gene for factor VIII was inserted into a retrovirus, a harmless virus that splices its own genes into the DNA of a host cell.

3. Hemophiliacs would be given the gene to make their own supply of factor VIII.

-44- Debugging the Blood

1. Donated blood is routinely tested for syphilis, hepatitis B and C, two types of HIV, two types of human T cell leukemia virus and other infectious agents.

2. Infectious viruses and bacteria cannot be destroyed in donated blood without destroying the blood itself.

3. Getting accurate medical histories and reporting cases of transfusion-transmitted diseases can help to identify unsuitable donors before they donate.

-45- Estrogen and Your Arteries

1. Estrogen levels are highest during the reproductive years and for women who take hormone replacement therapy during menopause.

2. Women who took estrogen had significantly higher levels of high-density lipoprotein which helps keep cholesterol from being deposited.

3. Estrogen may improve blood flow and prevent blood vessel spasms.

-46- The New Skinny on Fat

1. It turns down appetite and speeds up energy use causing animals to burn more fat.

2. Ob protein cannot be taken orally as it would be broken down during digestion.

3. Some forms of genetic obesity do not recognize the Ob protein.

-47- Diet and Breast Cancer

1. According to the NRC, fat should be restricted to 30% of the diet.

2. High fiber intake may reduce levels of circulating estrogens which can promote the growth of breast tumors.

3. Isoflavinoids

-48- Aspirin

1. Aspirin is the common name for the drug acetylsalicylic acid.

2. Aspirin inhibits the production of the chemical messenger prostaglandin.

3. Aspirin is being investigated for its role in preventing the blood clots responsible for heart attacks, preventing the severity of migraine headaches, improving circulation to the gums, lowering the risk of colorectal cancer and in controlling high blood pressure during pregnancy.

PART SEVEN: CONTROVERSIES AND ETHICAL ISSUES

-49- Cloning Human Embryos

1. They used the word "cloning" when describing their research and this description is very different from the popular notion of cloning.

2. Undifferentiated cells are easy to reproduce but adult cells are specialized and cannot be reproduced.

3. They performed the experiment to stir up an ethical debate on the value of cloning human embryos.

-50- The Future of the Fetal Tissue Bank

1. A fetal tissue bank would collect, process and distribute tissue throughout the country for tissue transplantation.

2. Elective abortions are more plentiful than tissue from spontaneous abortions and ectopic pregnancies.

3. Tissue should be tested rigorously to reduce infective and genetic risks to the recipient.

-51- The Mother of all Blood Cells

1. Stem cells are hematopoietic cells capable of generating an endless supply of other blood constituents like red cells, white cells and platelets.

2. He combined a number of different purification techniques to achieve a nearly pure batch of stem cells.

3. Stem cell therapies could result in new treatments for cancer, AIDS, leukemias and other inherited diseases.

-52- Researchers Face Delay in Supplies of RU486

1. RU486 blocks the action of the hormone progesterone and terminates pregnancy.

2. The most important potential uses of RU486 are in the treatment of breast cancer and brain tumors.

3. No manufacturers have been named due to the political opposition to RU486.